Methods in Molecular Biology

Series Editor
John M. Walker
School of Life and Medical Sciences
University of Hertfordshire
Hatfield, Hertfordshire, UK

For further volumes:
http://www.springer.com/series/7651

For over 35 years, biological scientists have come to rely on the research protocols and methodologies in the critically acclaimed *Methods in Molecular Biology* series. The series was the first to introduce the step-by-step protocols approach that has become the standard in all biomedical protocol publishing. Each protocol is provided in readily-reproducible step-by-step fashion, opening with an introductory overview, a list of the materials and reagents needed to complete the experiment, and followed by a detailed procedure that is supported with a helpful notes section offering tips and tricks of the trade as well as troubleshooting advice. These hallmark features were introduced by series editor Dr. John Walker and constitute the key ingredient in each and every volume of the *Methods in Molecular Biology* series. Tested and trusted, comprehensive and reliable, all protocols from the series are indexed in PubMed.

Biomimetic Sensing

Methods and Protocols

Edited by

Jessica E. Fitzgerald

Departments of Bioengineering and Chemical Engineering, Northeastern University, Boston, MA, USA

Hicham Fenniri

Departments of Chemical Engineering, Bioengineering, Chemistry and Chemical Biology, Northeastern University, Boston, MA, USA

Humana Press

Editors
Jessica E. Fitzgerald
Departments of Bioengineering
and Chemical Engineering
Northeastern University
Boston, MA, USA

Hicham Fenniri
Departments of Chemical Engineering, Bioengineering,
Chemistry and Chemical Biology
Northeastern University
Boston, MA, USA

ISSN 1064-3745 ISSN 1940-6029 (electronic)
Methods in Molecular Biology
ISBN 978-1-4939-9615-5 ISBN 978-1-4939-9616-2 (eBook)
https://doi.org/10.1007/978-1-4939-9616-2

Cover illustration: Barcoded polymer-based cross-reactive sensor array response to a mixture of analytes. The different colors illustrate different responses from the sensory elements.

This Humana imprint is published by the registered company Springer Science+Business Media, LLC, part of Springer Nature.
The registered company address is: 233 Spring Street, New York, NY 10013, U.S.A.

Preface

The Gap in Technology: Fast, Facile, and Quantifiable Detection of Analytes in Vapor and Liquid [1, 2]

Detection of analytes for vapor and liquid deconvolution has been performed in a variety of fields, including nutrition, toxicology, biomedicine, and chemistry. The gold standard for vapor sensing is gas chromatography-mass spectrometry (GC-MS), which is a quantitative sample analysis that provides both the type and amount of analytes present in a sample, usually in the form of volatile organic compounds (VOCs). While it is advantageous to know which specific VOCs are present in a sample, GC-MS is not practical for widespread use in vapor sensing because it requires both specialized, expensive equipment and highly trained personnel for operation. In addition, some relevant data may be lost due to vapor pre-concentration and sampling techniques. Similarly, the most common liquid sensing techniques require sample labeling, or tagging, before performing an assay. This sensing technique is limited because each analyte requires a specific label (making multiplexed sensing difficult), the analytes in the sample must be known beforehand so that the correct label can be selected, and nonspecific binding may take place, affecting the accuracy of the measured analyte concentration. As sensing needs to continue to expand and develop, there remains a need for methods that enable fast, facile, and accurate sensing at a low cost. To meet these needs, many researchers have looked to the mammalian olfactory system as a model for label-free, multiplexed sensing, developing platforms that are easy to use and produce and that can be used in a wide variety of applications.

The Mammalian Olfactory System as an Optimal Model for Multiplexed Sensing

There are about 1000 genes that encode olfactory receptors (ORs), and each OR has multiple sites for odorant binding, enabling the detection of more than one odorant for each OR, a characteristic called cross-reactivity. Different combinations of activated receptors make up unique signaling codes, or "fingerprints," for specific odorants, making it possible to distinguish between thousands. This sensing platform has inspired researchers over the past several decades to develop sensing devices that are cross-reactive and accurate and have multiplexing capabilities. These biomimetic devices are called "electronic/artificial noses" (e-noses) or e-tongues to detect certain analytes present in both vapors and liquids, respectively. E-devices have proved to be successful in a broad range of scientific and engineering fields, providing cost-effective, minimally invasive (in the case of clinical use), and highly accurate vapor and liquid component analysis. In this book, we highlight the potential of e-device technology to serve as a successful platform for multiplexed sensing, along with the methods for device fabrication, calibration, and assays in multiple applications. The subsequent sections describe e-device sensing platforms, explore their use, and outline existing limitations and future directions in device development.

e-Device History and Applications

The first use of the term "electronic nose" was at a conference in 1987, and the first conference dedicated specifically to artificial olfaction was in 1989. Gardner and Bartlett originally defined an "electronic nose" as follows: "*an instrument, which comprises of an array of electronic-chemical sensors with partial specificity and an appropriate pattern recognition system, capable of recognising simple or complex odours.*" The first devices that utilized this technology were comprised of sets of distinct active materials, each connected to its own signal transduction channel. When passing an analyte vapor over the sensor array, activated sensors would transmit an electrical signal to a processor, which would then alert the user of the analytes present by cross-referencing the list of activated sensors with a database of known analyte profiles. The earliest designs employed metal oxide semiconducting field effect transistors (MOSFETs) as electrical sensors to detect gases such as NO_2. This was based on the principle that the conductivity of semiconductor metals changes upon variance in the atmospheric gas surrounding the sensor. MOSFETs are usually constructed from a SiO_2 insulating layer, with a semiconductor metal deposited on top as the gate in the circuit. A voltage is applied to maintain a constant current, and as the gas adsorbs onto the gate, the conductance of the FET changes, thereby causing the voltage to change. e-devices are able to differentiate analytes via "fingerprint" outputs; that is, each sample's analyte profile (comprised of a unique mixture of analyte types and relative concentrations) produces a unique response pattern from the sensor array, enabling sample differentiation. New sample fingerprints are compared to known sample fingerprints via data analysis (in most cases multivariate data analysis such as principle component analysis), which enables pattern recognition, clustering, and classification of the unknown analyte sample.

This fingerprint sensing method has been implemented in a number of fields, including food science, quality control, drug testing, contamination detection, defense efforts (e.g., explosive detection), and medical diagnosis. Some of the most pioneering work has been done in medical diagnostics via exhaled breath analysis. Many metabolites, or metabolic by-products, have been identified and correlated with specific diseases; some of these can even be a good indicator of disease progression. Many of these metabolites are volatile organic compounds (VOCs), which are small molecules that enter exhaled breath through gas exchange at the alveolar-capillary membrane of the respiratory tract. While the VOCs produced in each disease are thought to be primarily from oxidative stress, the subsequent effect of each disease on the body is unique and leads to the production of disease-specific VOC profiles. A reliable, noninvasive device capable of detecting subtle molecular changes and differences can be leveraged to implement this personalized medicine approach. The e-devices included in this work have much potential for implementation in these fields and others, as their sensitivity and specificity can be tuned toward specific analytes of interest.

e-Device Advances in Technological Development

Within the past several years, researchers have taken advantage of rapid technological advances to expand and improve e-device sensing platforms and sensor materials, signal transduction mechanisms, and data processing for pattern recognition and analyte identification. Moreover, as more became known about the physiology of the mammalian olfactory system, sensors further advanced to take advantage of these new findings, from incorporating sensors that were able to detect multiple analytes to the development of a biomimetic flow chamber to enhance analyte detection at low levels. This allowed for a greater number of analytes to be recognized using a smaller array of multi-selective sensors. Within the last two decades, electronic tongues for liquid detection have also been developed, mimicking the olfactory signaling pathways.

The types of sensors employed for both vapor and liquid samples in these devices vary, including gravimetric, or mass, electrochemical, and optical sensors, allowing characterization of analytes based on mass, electrical properties (e.g., conductance impedance), and electron/photon interactive properties, respectively. Gravimetric sensors are either piezoelectric (PZ) crystals or microcantilevers, which have a specific resonant frequency. On binding with an analyte, the resonant frequency of the sensor drops in proportion to the added mass, due to either viscoelastic or gravimetric effects. Electrochemical e-devices are comprised of an electronic circuit connected to a network of sensory materials—most commonly conductive polymers or metal oxides—that provide an electrical response on binding with a specific known analyte. This response is characterized by monitoring sensor conductivity, resistivity, or voltage change during vapor exposure. Finally, optical sensors work by displaying a shift in the emission or absorption of different types of electromagnetic radiation on binding with a desired analyte. There are two popular means of optical detection: fluorescent sensors, which fluoresce upon analyte binding, and colorimetric sensors, which display a visible color change upon analyte binding.

Although gravimetric sensors have proved to be successful, there are many limitations with this device setup. Inaccuracies due to subtle changes in surface coating, humidity, or temperature necessitate frequent calibrations, which is unfortunately delicate and time-consuming. e-devices that employ optical and electrochemical sensors have shown much promise as they provide an easier and more cost-effective way of identifying analytes in vapor while maintaining accuracy. Optical sensors offer significant benefits compared with those mentioned above since they can provide multiple complex data types simultaneously, including changes in intensity, fluorescence lifetime, wavelength, and spectral shape. This approach increases the ratio of recognizable analytes to the number of sensors used. This work mainly focuses on optical (Chapters 1–10) and electrochemical (Chapters 11–13) methods of sensing, as these are at the forefront of e-device technology, being more stable and reliable than their gravimetric counterparts. We have also included a chapter on a cutting-edge mechanochemical sensing method using folded DNA origami structures (Chapter 14) that have been demonstrated to have a limit of detection down to the single molecule level. Finally, we highlight here some cutting-edge methods to optimize e-device data and technology via drift correction and calibration (Chapter 15) and computer modeling of sensor output for material optimization (Chapter 16).

Optical Sensor Arrays for e-Devices

Optical sensors in e-device systems have shown much promise to provide a facile, cost-effective, and accurate way of identifying analytes in vapor and liquid samples. Optical sensors display a shift in emission or absorption of different types of radiations upon analyte binding. The two most popular means of detection are spectroscopic and colorimetric sensing. In this work, we highlight the methods and applications of both means of detection. The spectroscopic methods employ Raman spectroscopy (Chapter 1), interferometry (Chapter 2), mass spectrometry (Chapter 3), fluorescence microscopy (Chapter 5), and surface plasmon resonance (Chapters 8 and 10). The colorimetric sensors herein are produced in several ways, including microchip fabrication and photolithography (Chapters 4, 6, and 7). As mentioned previously, optical sensors offer significant benefits compared with gravimetric sensors; they can provide multiple complex data types simultaneously, including changes in intensity, fluorescence and colorimetric lifetime, wavelength, and spectral shape. This increases the ratio of recognizable analytes to the number of sensors used. In addition, microsphere optical arrays such as those developed by Walt et al. and presented in Chapter 1 provide an advantage over other multisensor systems because billions of beads that produce an identical response can be made simultaneously, compared with many sensors for which the fabrication process is tedious. Each type of bead has a distinct, intrinsic response to the samples presented, which eliminates the need for additional encoding for bead identification.

Electrochemical Sensor Arrays for e-Devices

Electrochemical sensing e-devices use an electronic circuit connected to a network of sensory materials that provide an electrical response upon binding with a specific known analyte. While the most common electrochemical sensing platform is the MOSFET devices, FET sensors have also been developed that incorporate organic material such as DNA-decorated graphene FETs (Chapter 13). Another popular, recent sensory material is conductive polymer (Chapters 11 and 12). Chapter 11 describes a method for layer-by-layer deposition of conducting polymers in a microfluidic channel as an electronic tongue. The conducting polymers described in Chapter 12 are electropolymerized in the presence of the target molecule or template which is then removed after polymerization to create molecularly imprinted polymers. In both cases, the polymer sensors are then placed in an electrical circuit and act as resistors, reflecting a decrease in their conductance (or an increase in impedance) upon binding with the analytes in the sample. This decrease in conductance is most likely due to polymer swelling upon analyte binding—as the polymer swells, gaps between polymer chains increase, lowering conductivity.

Remaining Challenges and Future Outlook for e-Device Implementation

As e-device implementation continues to grow in breadth, there are certain limiting factors that must be addressed. First, there remains a lack of standards for sample collection, both environmental (ambient air, water) and medical (exhaled breath, saliva). When developing a sampling method, it is important to optimize the collection, preparation, and storage method to maximize analyte detection without denaturing or altering the chemical profile of the sample. For example, collecting *alveolar* breath, the second phase of the breathing cycle, requires a sampling method that minimizes VOC interference from ambient air while capturing the alveolar air from a patient who is breathing steadily at a set velocity. The sample storage material and time of storage also affect analyte recovery. Even after obtaining an ideal sample, e-device performance accuracy may be limited by extrinsic factors such as humidity and temperature or intrinsic factors such as sensor drift and instrumentation errors. Additionally, e-device analyte fingerprint analysis via pattern recognition requires complex data analysis, which currently limits the widespread implementation of these devices.

Finally, though preliminary studies with e-devices have been largely successful, their reproducibility is limited because methods must be optimized de novo for each specific application. Standards need to be developed from statistical analysis of device performance and should include thresholds for success in areas such as response reproducibility, specificity, and sensitivity. In developing these standards, it is also important to consider the ultimate goal of the device. For example, if the goal is simply to diagnose and classify a sample, such as one that correlates with a specific disease, selectivity is more important than sensitivity; however, if the goal is to monitor analyte profile change over time, such as with disease progression, sensitivity to slight variations in analyte concentrations is of great importance.

Though these limitations currently serve as a bottleneck for widespread e-device implementation, researchers continue to work diligently to develop methods that will circumvent and overcome them. For example, Chapter 15 includes a detailed method for device calibration that can be applied to a wide variety of sensing mechanisms and e-devices. Additionally, complementary methods, such as computer modeling of analyte-sensor interaction presented in Chapter 16, serve as a way to better predict sensor response to specific analytes. This can then be leveraged to produce devices that are highly tuned toward a specific analyte or group of analytes, i.e., much improved sensitivity and selectivity. As the technology for e-device sensing continues to trend toward portable, accurate, and easy-to-use platforms, they have great potential to be implemented wherever analyte detection is required. Indeed, they may be able to reduce the need for highly specialized, expensive equipment and personnel. For developing countries in particular, e-devices that are simultaneously inexpensive may soon be able to take the place of highly specialized equipment in fields such as defense efforts and explosive detection, water and food contamination, and personalized medicine.

Boston, MA, USA

Jessica E. Fitzgerald
Hicham Fenniri

References

1. Fitzgerald JE, Bui ETH, Simon NM, Fenniri H (2016) *Trends Biotechnol.* 1
2. Fitzgerald JE, Fenniri H (2016) *RSC Adv.* 6(84): 80468

Contents

Contributors

NAGEH K. ALLAM • *Energy Materials Laboratory, School of Sciences and Engineering, The American University in Cairo, New Cairo, Egypt*

MARIA L. BRAUNGER • *Department of Applied Physics, "Gleb Wataghin" Institute of Physics (IFGW), University of Campinas—UNICAMP, Campinas, SP, Brazil*

COLLIN J. BRIGHT • *U.S Army Combat Capabilities Development Command, C5ISR Center, Fort Belvoir, VA, USA*

OLALLA CALVO-LOZANO • *Nanobiosensors and Bioanalytical Applications Group (NanoB2A), Catalan Institute of Nanoscience and Nanotechnology (ICN2), CSIC and BIST, Barcelona, Spain; Networking Center on Bioengineering, Biomaterials and Nanomedicine (CIBER-BBN), Barcelona, Spain*

DANIEL CITTERIO • *Faculty of Science and Technology, Department of Applied Chemistry, Keio University, Yokohama, Japan*

CRISTIANE M. DAIKUZONO • *Centro de Ciências e Tecnologias para Sustentabilidade (CCTS), Universidade Federal de São Carlos—UFSCar, Sorocaba, SP, Brazil*

ADIL DENIZLI • *Department of Chemistry, Hacettepe University, Ankara, Turkey*

MARI C. ESTÉVEZ • *Nanobiosensors and Bioanalytical Applications Group (NanoB2A), Catalan Institute of Nanoscience and Nanotechnology (ICN2), CSIC and BIST, Barcelona, Spain; Networking Center on Bioengineering, Biomaterials and Nanomedicine (CIBER-BBN), Barcelona, Spain*

HICHAM FENNIRI • *Departments of Chemical Engineering, Bioengineering, Chemistry and Chemical Biology, Northeastern University, Boston, MA, USA; Department of Bioengineering, Northeastern University, Boston, MA, USA*

JESSICA E. FITZGERALD • *Department of Bioengineering and Department on Chemical Engineering, Northeastern University, Boston, MA, USA*

M. TERESA S. R. GOMES • *Chemistry Department, University of Aveiro, Aveiro, Portugal; CESAM, University of Aveiro, Aveiro, Portugal*

VERA L. V. GRANADO • *Chemistry Department, University of Aveiro, Aveiro, Portugal*

YUMA IKEDA • *Faculty of Science and Technology, Department of Applied Chemistry, Keio University, Yokohama, Japan*

A. T. CHARLIE JOHNSON • *Department of Physics and Astronomy, University of Pennsylvania, Philadelphia, PA, USA*

SAGUN JONCHHE • *Department of Chemistry and Biochemistry, Kent State University, Kent, OH, USA*

DMITRY KIRSANOV • *Institute of Chemistry, St. Petersburg State University, St. Petersburg, Russia; Laboratory of Artificial Sensory Systems, ITMO University, St. Petersburg, Russia*

LAURA M. LECHUGA • *Nanobiosensors and Bioanalytical Applications Group (NanoB2A), Catalan Institute of Nanoscience and Nanotechnology (ICN2), CSIC and BIST, Barcelona, Spain; Networking Center on Bioengineering, Biomaterials and Nanomedicine (CIBER-BBN), Barcelona, Spain*

ANDREY LEGIN • *Institute of Chemistry, St. Petersburg State University, St. Petersburg, Russia; Laboratory of Artificial Sensory Systems, ITMO University, St. Petersburg, Russia*

FENGYU LI • *Key Laboratory of Green Printing, Institute of Chemistry, Chinese Academy of Sciences (ICCAS), Beijing Engineering Research Center of Nanomaterials for Green*

Printing Technology, Beijing National Laboratory for Molecular Sciences (BNLMS), Beijing, P. R. China

NATHAN C. LINDQUIST • *Department of Physics and Engineering, Bethel University, St. Paul, MN, USA*

LARISA LVOVA • *Laboratory of Artificial Sensory Systems, ITMO University, St. Petersburg, Russia; Department of Chemical Science and Technologies, University "Tor Vergata", Rome, Italy*

HANBIN MAO • *Department of Chemistry and Biochemistry, Kent State University, Kent, OH, USA*

TAKAMICHI NAKAMOTO • *Institute of Innovative Research, Tokyo Institute of Technology, Yokohama, Kanagawa, Japan*

JAMES W. NEEDHAM • *InBios International Inc., Seattle, WA, USA*

YUJI NOZAKI • *Institute of Innovative Research, Tokyo Institute of Technology, Yokohama, Kanagawa, Japan*

VITALY PANCHUK • *Institute of Chemistry, St. Petersburg State University, St. Petersburg, Russia; Laboratory of Artificial Sensory Systems, ITMO University, St. Petersburg, Russia*

ENELIA C. PELÁEZ • *Nanobiosensors and Bioanalytical Applications Group (NanoB2A), Catalan Institute of Nanoscience and Nanotechnology (ICN2), CSIC and BIST, Barcelona, Spain; Networking Center on Bioengineering, Biomaterials and Nanomedicine (CIBER-BBN), Barcelona, Spain*

JINGLEI PING • *Department of Mechanical and Industrial Engineering, University of Massachusetts Amherst, Amherst, MA, USA*

ALEJANDRO PORTELA • *Nanobiosensors and Bioanalytical Applications Group (NanoB2A), Catalan Institute of Nanoscience and Nanotechnology (ICN2), CSIC and BIST, Barcelona, Spain; Networking Center on Bioengineering, Biomaterials and Nanomedicine (CIBER-BBN), Barcelona, Spain*

MENG QIN • *Key Laboratory of Green Printing, Institute of Chemistry, Chinese Academy of Sciences (ICCAS), Beijing Engineering Research Center of Nanomaterials for Green Printing Technology, Beijing National Laboratory for Molecular Sciences (BNLMS), Beijing, P. R. China*

ISABEL S. RICH • *Department of Physics and Engineering, Bethel University, St. Paul, MN, USA*

ANTONIO RIUL JR • *Department of Applied Physics, "Gleb Wataghin" Institute of Physics (IFGW), University of Campinas—UNICAMP, Campinas, SP, Brazil*

ALISA RUDNITSKAYA • *Chemistry Department, University of Aveiro, Aveiro, Portugal; CESAM, University of Aveiro, Aveiro, Portugal*

VINCENT P. SCHNEE • *U.S Army Combat Capabilities Development Command, C5ISR Center, Fort Belvoir, VA, USA*

SPENCER T. SEILER • *Department of Physics and Engineering, Bethel University, St. Paul, MN, USA*

VALENTIN SEMENOV • *Institute of Chemistry, St. Petersburg State University, St. Petersburg, Russia*

GÜLSU ŞENER • *Department of Chemistry, Hacettepe University, Ankara, Turkey*

ICELL M. SHARAFELDIN • *Energy Materials Laboratory, School of Sciences and Engineering, The American University in Cairo, New Cairo, Egypt*

HIROYUKI SHIBATA • *Faculty of Science and Technology, Department of Applied Chemistry, Keio University, Yokohama, Japan*

YANLIN SONG • *Key Laboratory of Green Printing, Institute of Chemistry, Chinese Academy of Sciences (ICCAS), Beijing Engineering Research Center of Nanomaterials for Green Printing Technology, Beijing National Laboratory for Molecular Sciences (BNLMS), Beijing, P. R. China*
M. SELIM ÜNLÜ • *Department of Biomedical Engineering, Boston University, Boston, MA, USA; Department of Electrical and Computer Engineering, Boston University, Boston, MA, USA*
NESE LORTLAR ÜNLÜ • *Department of Biomedical Engineering, Boston University, Boston, MA, USA*
CELALETTIN YURDAKUL • *Department of Electrical and Computer Engineering, Boston University, Boston, MA, USA*

Chapter 1

Cross-Reactive, Self-Encoded Polymer Film Arrays for Sensor Applications

Jessica E. Fitzgerald and Hicham Fenniri

Abstract

The development of chemical sensors continues to be an active area of research, especially the development of a practical electronic nose. Here, we present a spectroscopic chemical sensor based on an array of 64 self-encoded polymer films deposited on a microfabricated silicon substrate. The polymer arrays were analyzed by FTIR and Raman spectroscopy before and after exposure to a series of organic volatiles to monitor changes in their vibrational fingerprints. We show here that the spectroscopic changes of self-encoded polymer films can be used to distinguish between volatile organic analytes. Changes induced in the sensor arrays by the analyte vapor were denoted by a spectroscopic response of the self-encoded polymer sensors and transformed into a response pattern by multivariate data analysis using partial least squares regression. The results indicated that the polymer sensors provide a unique and reproducible pattern for each analyte vapor and can potentially be used in the fabrication of a novel electronic nose device.

Key words Barcoded polymer, Sensor array, Raman spectroscopy, Multivariate data analysis

1 Introduction

Biomimetic engineering is the application of biological principles to the design of artificial devices or systems. For many years, scientists and engineers have recognized the power of naturally occurring systems and their ability to guide the development of certain technologies. One example of the biomimetic approach is a device known as the electronic or artificial nose (e-nose) [1]. Gardner and Bartlett define an e-nose as "an instrument which comprises an array of electronic chemical sensors with partial specificity and an appropriate pattern recognition system, capable of recognizing simple or complex odors" [2]. Applications of e-noses include food quality control [3], pollution monitoring [4], medical diagnosis [5], and landmine detection [6], among others. In this rapidly growing field, arrays of semi-selective chemical sensors are the main component of the devices. These arrays are coupled to pattern recognition software, paralleling the biological olfactory system in

Jessica E. Fitzgerald and Hicham Fenniri (eds.), *Biomimetic Sensing: Methods and Protocols*, Methods in Molecular Biology, vol. 2027, https://doi.org/10.1007/978-1-4939-9616-2_1, © Springer Science+Business Media, LLC, part of Springer Nature 2019

which semi-selective olfactory receptors are combined with higher order neural processing [7]. It is thought that there are millions of receptor cells made up of approximately 1000 different receptor types in the mammalian olfactory epithelium; while none of the receptors are thought to be particularly sensitive to one specific analyte, it is likely that their combined signal leads to the highly sensitive and discriminative sense of smell [8, 9], Similarly, in e-nose system architecture, no individual detector is highly selective toward an individual analyte, as would be the case in the traditional "lock and key" approach to chemical sensing. Instead, each detector responds to many analytes with varying degrees of intensity, creating a unique response pattern for each analyte, a principle known as cross-reactivity. The resulting odor signature from the array is used to classify, and in some cases quantify, the analyte of concern.

E-noses are primarily classified by the sensor-transduction mechanism. E-nose transducers reported to date include semiconducting metal oxide [10], conducting polymer films [11], acoustic wave devices [12], electrochemical systems [13], carbon-black loaded polymer film chemoresistors [14], and optical transducers with immobilized dyes such as Nile Red or various metalloporphyrins [15, 16]. Among these approaches, most array sensors employing polymers have produced high selectivity, taking advantage of properties such as polarity, swelling, conductivity, and sorption. Regardless of the transduction mechanism of the sensor array, a larger number of unique sensors increases the quantity and accuracy of the data gathered, thus producing a more complex and specific pattern for improved analyte identification and classification.

The research and development of sensor arrays often involves the measurement and analysis of a large number of samples. This can be very laborious and time consuming because there are many variables that influence the performance, sensitivity, and precision of the sensors. Therefore, there is currently a significant interest in the development of sensor arrays with high selectivity, sensitivity, reproducibility, and high-throughput capability for applications in disease diagnosis from patients' vapor/breath analysis.

Our group has recently reported on a new class of resins prepared from spectroscopically active styrene monomers, the combination of which produces polymers wherein unique vibrational fingerprints are associated with each polymer. The spectrum from each polymer can then be converted into a barcode in which the position of each bar matches a peak wavenumber in the spectrum. We have previously demonstrated that each barcoded resin (BCR) can be selectively recognized and classified by the unique self-encoded spectra that are produced from the BCRs' composition. BCRs have previously been proposed for use in the deconvolution strategies of resin-supported combinatorial libraries [17].

In this chapter, we describe both the fabrication of polymer film sensor arrays using BCRs and the testing of the arrays for the detection and differentiation of volatile organic compounds (VOCs) in vapor [18] and liquid [19] phases. Polymer sensor arrays were fabricated by depositing a selection of BCRs into micromachined wells on silicon substrates. The analysis of the vibrational spectroscopic changes by partial least squares regression (PLS) efficiently captures the variances in the data set and allows discrimination of the VOCs. Contour plots and histogram graphs representing the changes in the composition matrix for each polymer sensor in the array indicated that the array responses to VOCs were sensitive and reproducible.

2 Materials

1. Suspension copolymerization:
   (a) Benzoyl peroxide (BPO).
   (b) Poly(vinylalcohol) (PVA).
   (c) 80% Divinylbenzene (DVB).
   (d) Chloromethylstyrene.
   (e) Co-monomers, distilled under reduced pressure to remove the stabilizers and then stored at +4 °C: styrene, 2,5-dimethylstyrene, 4-methylstyrene, 2,4-dimethylstyrene, 4-t-butylstyrene, 3-methylstyrene (*see* **Note 1**).
   (f) The following solvents, distilled prior to use: *N,N*-dimethylformamide (DMF), dichloromethane (DCM), and toluene were distilled from CaH_2; methanol and ethanol from Mg, and tetrahydrofuran (THF) from Na/benzophenone.
   (g) Polypropylene woven mesh (0.0030″ opening width).
   (h) An IKA–RW20 mechanical motor.
   (i) The reaction vessels and impellers were designed according to the literature and shown in Fig. 1 [20]. A piece of Pyrex tube of *ca.* 70 mm external diameter and 3–4 mm thickness was placed in a glass working lathe, and by the use of a specially made metal jig four equally spaced indentations (baffles) were pressed along its length. After annealing, the extension of a B40 Pyrex socket was joined to one end of the tube, and two B19 sockets were then attached to either side of the B40 socket. A B10 socket was precisely positioned at one side of one of the indentations at the head of the vessel. The other end of the tube was then smoothly rounded off and the whole vessel annealed.
   (j) Standard Morton flask (ChemGlass).

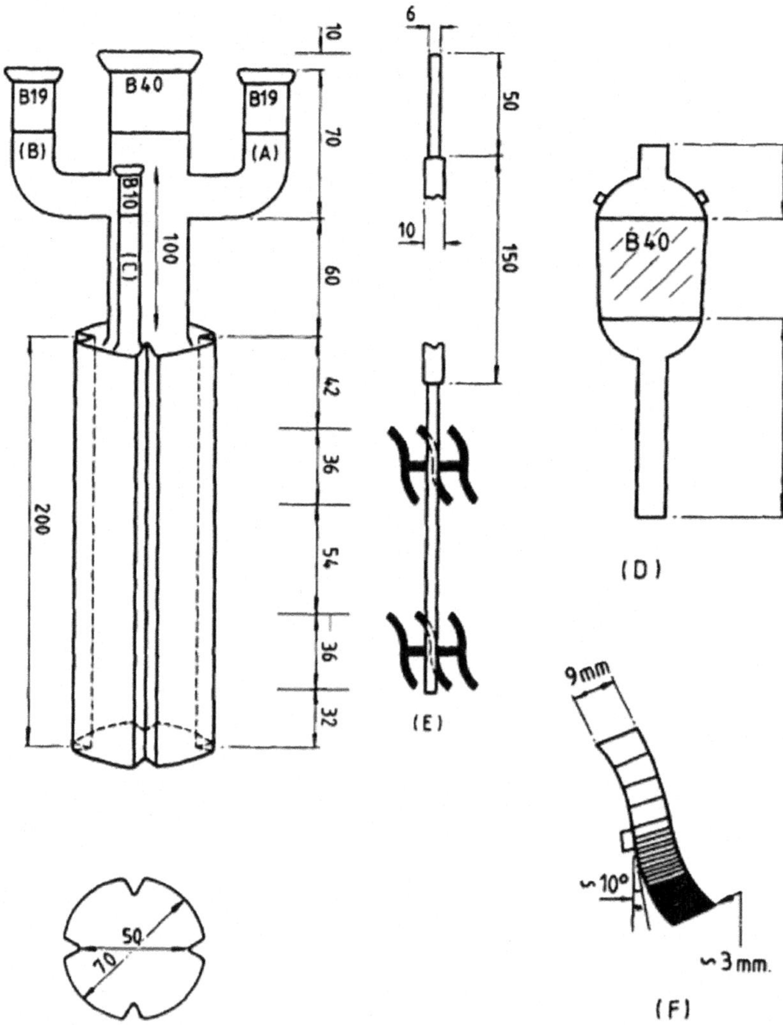

Fig. 1 Suspension polymerization apparatus. (**a**) and (**b**) are accession points for a reflux condenser and nitrogen inlet, (**c**) is a sampling area, (**d**) is a stirrer guide, (**e**) is the stirrer with its position indicated for full-scale operation. (**f**) is an expanded drawing of a single stirrer blade indicating the curvature at both ends and the position and angle of attachment to the stirrer rod. All dimensions are in mm. Used with permission from [20]

2. Substrate array fabrication:
   (a) Clean room.
   (b) Photolithography machinery.
   (c) Spin coater.
   (d) Silicon substrates.
   (e) Piranha cleaning solution.
   (f) HMDS primer.
   (g) AZ P4620 photoresistor.

(h) Mask aligner.

(i) Anisotropic dry reactive etching equipment.

(j) RF sputter coater.

3. Rubberized spatula.
4. Kimwipes.
5. Clinically relevant analyte vapors: acetone, pentane, propyl benzene, cumene, carbon disulfide (*see* **Note 2**).
6. Raman microspectrometer (*see* **Note 3**): laser excitation wavelength of 785 nm, laser power of *ca.* 40 mW, spectral range 500–1800 nm^{-1}.
7. Data processing: OriginPro 2016®, Microsoft Excel, and Unscrambler ® (Camo) software, or some other software with multivariate data analysis capabilities.

3 Methods

1. Suspension copolymerization (*see* **Note 4**) [20, 21]:

(a) Place deionized water (190 mL) and 10% (w/w) PVA/H_2O (4 g) in the reaction vessel equipped with a mechanical stirrer, condenser, and N_2 inlet. The reaction should be kept under N_2 atmosphere throughout the entire polymerization process.

(b) Add the organic solution composed of co-monomers (Fig. 2), DVB, chloromethylstyrene (for weight percent values of 25 different BCRs, *see* Table 1), and BPO (0.100 g) to the reaction vessel (*see* **Note 5**).

(c) Stir the mixture at a fixed speed (300–400 rpm, radius of stir blades *ca.* 50 mm) to produce the desired bead size distribution.

(d) Immerse the reactor in a preheated oil bath maintained at 80 °C.

(e) After 24 h, stop the motor, filter the beads formed, and wash with deionized water (dH_2O).

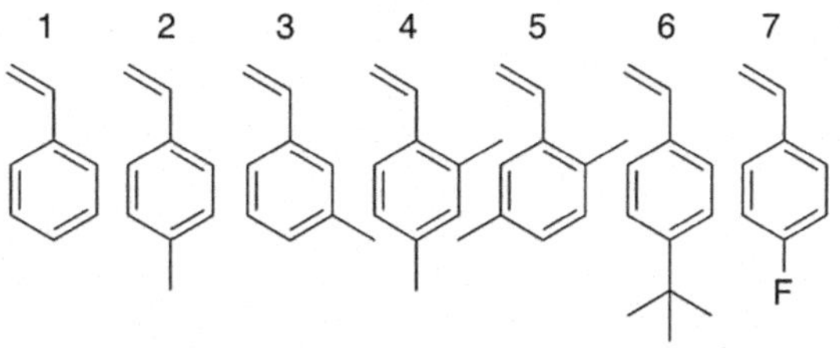

Fig. 2 Molecular structure of seven different monomers used for the fabrication of the barcoded copolymers. Used with permission from [18]

Table 1
Weight percent of styrene monomers used for the synthesis of 25 BCR sensors

BCR ID	Styrene	2,5-Dimethylstyrene	4-Methylstyrene	2,4-Dimethylstyrene	4-t-Butylstyrene	3-Methylstyrene
1	100	0	0	0	0	0
2	33	33	0	0	0	33
3	33	0	0	0	33	33
4	0	33	0	0	33	33
5	33	33	0	0	33	0
6	0	33	33	0	0	33
7	0	50	0	0	0	50
8	50	0	0	0	50	0
9	33	0	33	0	0	33
10	0	0	100	0	0	0
11	0	0	50	0	0	50
12	0	0	50	0	50	0
13	50	0	50	0	0	0
14	0	50	50	0	0	0
15	0	33	33	0	33	0
16	50	0	0	0	0	50
17	0	0	33	0	33	33
18	50	0	0	50	0	0
19	0	50	0	0	50	0
20	33	0	33	0	33	0
21	0	0	0	0	50	50
22	0	100	0	0	0	0
23	33	33	33	0	0	0
24	50	50	0	0	0	0
25	0	0	0	0	0	100

(f) Wash the beads with dH_2O and ethanol using a Soxhlet extractors (24 h each).

(g) Finally, sieve the beads and dry under vacuum.

2. Sensor array fabrication: Lithography and anisotropic dry reactive ion etching (RIE) (*see* **Note 6**).

(a) Piranha clean the silicon substrates and prime with HMDS primer.

(b) Spin-coat with AZ P4620 photoresists.

(c) Expose areas of the silicon by irradiating the film ($\lambda = 400$ nm) through a lithography mask.

(d) Etch the exposed silicon wafer with RIE using a combination of SF_6/C_2F_4, at pressures of 125 and 75 sccm, respectively. In this way, 10, 100, and 200 μm deep wells can be obtained.

(e) Coat the patterned wafer with a 200 μm thick copper film by RF sputtering.

3. Polymer array deposition (*see* **Notes** 7 and **8**):

(a) Deposit the polymer beads onto the micro-patterned silicon substrates by dusting an amount of beads on the surface of the silicon chip.

(b) Use a rubberized spatula to spread the powder in a circular motion over the chip to ensure the filling of the holes. SEM of the BCR arrays (Fig. 3) clearly shows that by

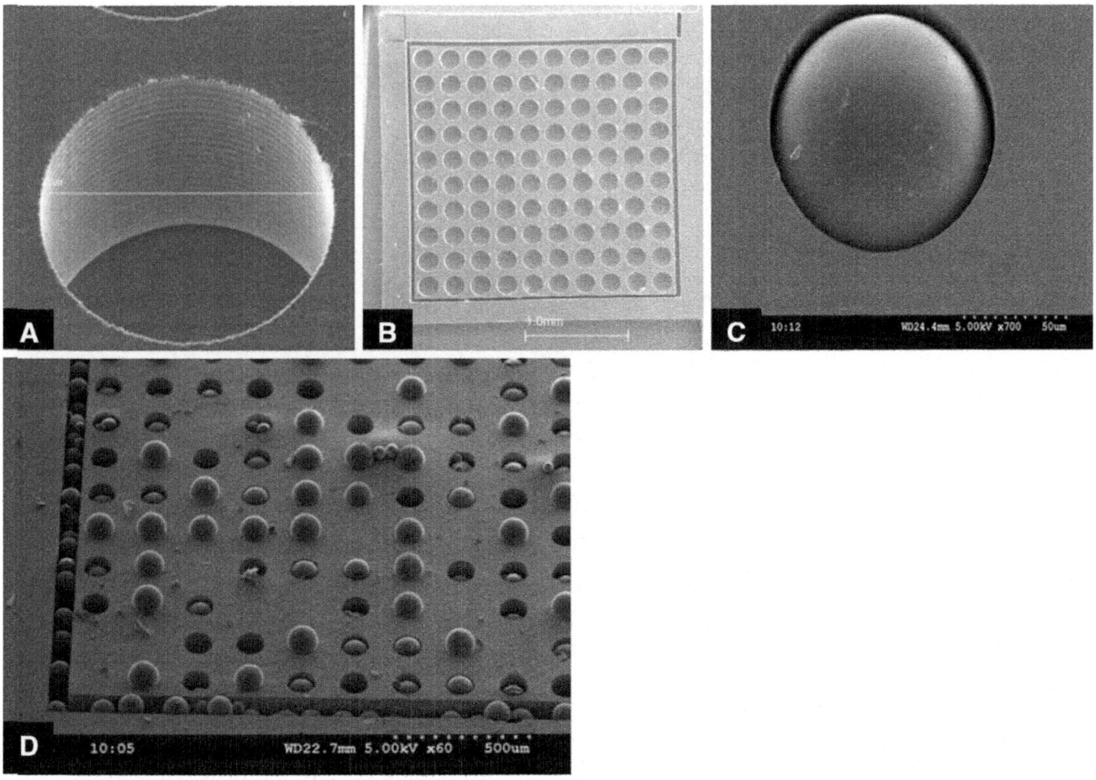

Fig. 3 SEM images of etched micro-well (**a**), full substrate (**b**), deposited BCR bead (**c**), and complete sensor array (**d**)

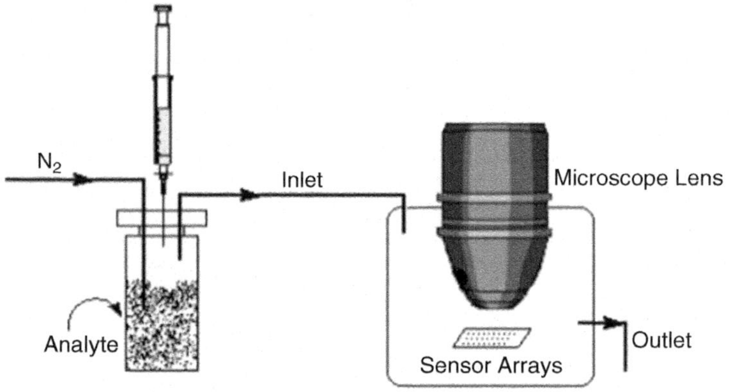

Fig. 4 Schematic diagram of the analyte delivery setup and analyzing system for recording response of the sensor array to the application of analyte (general setup of the experiment). Exposure to a given analyte vapor induces changes in the (IR or Raman) spectrum that are recorded. Used with permission from [7]

selecting the right bead/well size combination, an ideal one well/one bead distribution can be achieved.

   (c) Remove excess beads by wiping with a kimwipe.

4. Analyte vapor exposure measurements:
   (a) First, place the polymer sensor array inside the gas flow cell and collect Raman spectra prior to analyte vapor exposure.
   (b) Pass the saturated analyte vapor, generated by slowly streaming nitrogen through a vial containing the liquid of interest, through the gas flow cell containing the sensor arrays. An example of the vapor exposure setup is shown in Fig. 4.
   (c) After analyte vapor exposure, allow the sensing system to reach equilibrium before spectral acquisition or hyperspectral imaging. The system is determined to be at equilibrium when no further changes are recorded in the spectra.
5. Liquid analyte exposure measurements: A CRSA consisting of BP sensor elements patterned on a silicon chip mounted on a glass slide can also be operated as an artificial "tongue":
   (a) Leave a row of the sensor array wells empty (*i.e.*, bead-free) to test for Raman scattering emanating from the analytes at the working concentrations (0.25 mg/L).
   (b) Focus the laser (785 nm, 5.4 mW) onto the sample with a 5× objective (providing scattering areas of *ca.* 10 μm^2) and take scans before and after addition of the liquid analyte solutions, with acquisition times of 10 s. Run each spectrum in triplicate.

(c) Extract the collected Raman spectra of the polymers before and after exposure to analytes and import into data analysis software for pre-processing before multivariate data analysis.

6. Data processing: The preprocessing of the vibrational spectra involves baseline correction and normalization to compensate for the differences in intensities of the spectra. This procedure should be performed for each spectrum before and after VOC exposure:

(a) Import the spectra of the BP sensors before and after exposure into the multivariate data analysis software.

(b) Plot all raw spectra and filter to eliminate any interference from γ-rays.

(c) Normalize spectra to 1 and subtract the baseline using the second derivative function.

(d) Smooth the corrected spectra using a 5-point Savitzky-Golay algorithm, and take the first derivative of each spectrum to eliminate any further discrepancies in the baseline between spectra. Example before and after spectra of a BP sensor exposed to methanol is shown in Fig. 5.

(e) The spectral matrix is a collection of vectors for each BP, each with a unique spectral vector in ***n***-dimensional space, y_i, where each dimension represents a relative intensity for any given wavenumber (with ***n*** wavenumbers in the measured range). To compare the response of sensor arrays to the analytes, the vectors obtained from the spectra matrix before and after the analyte vapor treatment are considered the "reference" and "response" matrices, respectively. For a given trial, ***m*** BPs are used to construct

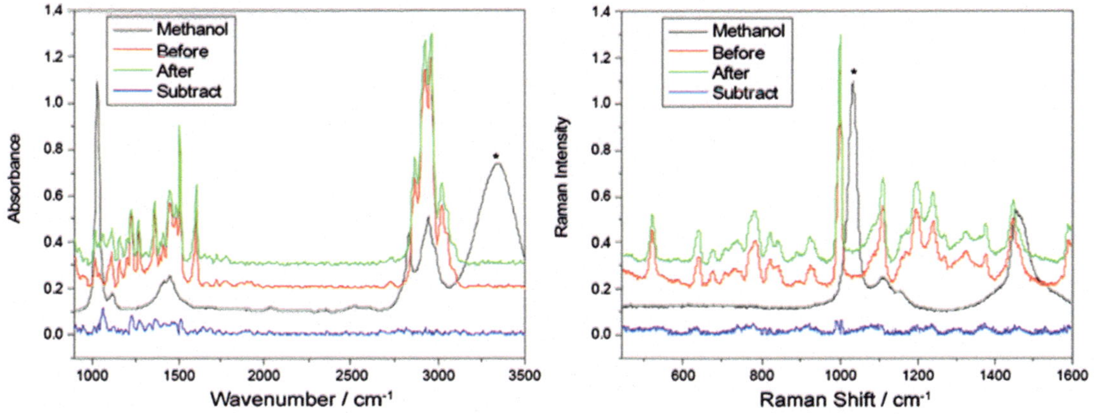

Fig. 5 FTIR (left) and Raman (right) of BP (3-methylstyrene/4-t-butylstyrene) before and after methanol vapor exposure. Methanol and difference spectra between "before" and "after" methanol vapor. Used with permission from [18]

a sensor array, and thus spectral data was represented as a spectral matrix of reference and response scans (**Y** and **Y**′, respectively) with dimensions $\boldsymbol{m} \times \boldsymbol{n}$.

(f) An angle map can then be generated, where $\boldsymbol{\theta}$ is defined as the angle between the reference and response vectors for each BP sensor. The angle between every $\boldsymbol{y_i}$ and $\boldsymbol{y'_i}$ was determined by the inner product of the two vectors:

$$cos\theta = \frac{|y_i * y_i'|}{|y_i||y_i'|} \tag{1}$$

(g) This gives $\boldsymbol{n^2}$ vector pairings or angles. From Eq. 1 it can be seen that if $\boldsymbol{y_i} = \boldsymbol{y'_i}$ then $\mathbf{cos}\ \boldsymbol{\theta} = \mathbf{1}$; thus $\boldsymbol{\theta} = \mathbf{0}$. This shows that the closer the response vectors are to the reference vectors, the smaller the angle between y_i and y'_i will be. Moreover, by comparing BP vectors before (reference) and after (response) the interaction with the analyte vapor, a response pattern could be generated for each analyte. Therefore, the angle reflects the degree of spectral change of each BP sensor due to the analyte vapor exposure. For example of $\boldsymbol{\theta}$ values from nine different BP sensors, see Fig. 6.

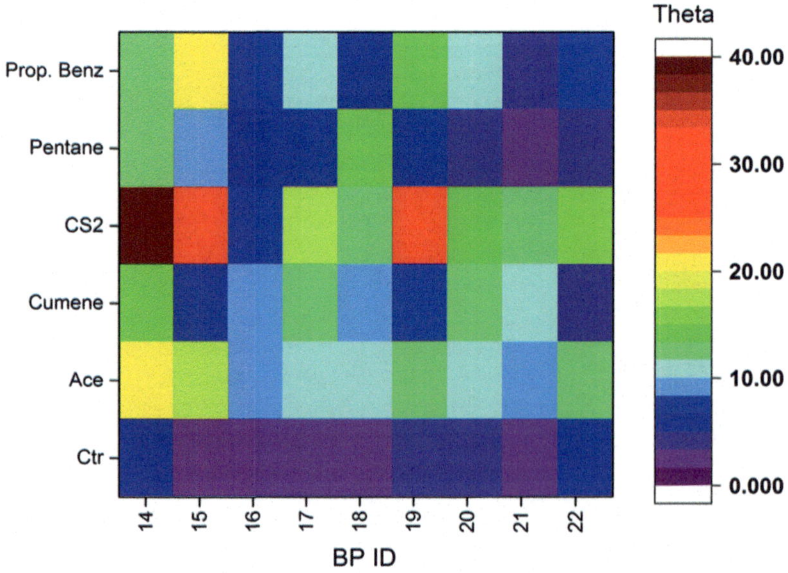

Fig. 6 Theta heat map of polymer/analyte interactions compared to the control (Ctr). Analytes studied: acetone (Ace), cumene, carbon disulfide (CS2), pentane, propyl benzene (Prop. Benz)

4 Notes

1. The monomers listed in Subheading 3 for polymerization are a sublibrary of the many monomers our lab has used to synthesize BCR beads. For a complete list of successfully used monomers for BCR bead synthesis, see reference [17]. The array can be tuned for both specificity and selectivity of a desired analyte by creating a library of BCR compositions and their corresponding response to each analyte. The beads with the most unique response (spectral shape change) and strongest response (intensity of change) to the desired analyte can be used to fabricate an array with highest specificity and sensitivity, respectfully.
2. The analytes chosen here are also a sublibrary of the volatile organic compounds that have been detected in the exhaled breath of diseased patients. These were selected both because of their chemical diversity and clinical relevance. The array can be used for other analytes of clinical relevance depending on the disease to be diagnosed/detected.
3. The BCR beads can also be excited at the 585 or 633 nm wavelength; however, we have found in our experiments that the 785 nm wavelength gives the best signal-to-noise ratio (lowest background fluorescence). Moreover, the BCRs also have unique Fourier transform infrared (FTIR) spectra [17, 22], so the analyte exposure experiments can also be performed using and FTIR spectrometer.
4. The main bead size fraction was in the range of 70–140 mesh (212–106 mm). Based on the yield of the suspension polymerization derived from the mass balance (>95%) we inferred that all the styrene monomers used were effectively incorporated in the beads. This conclusion was also inferred from FTIR and Raman spectroscopy [21].
5. The styrene monomers (Fig. 2) used to prepare the encoded polymers were chosen (a) based on their commercial availability, (b) to maximize spectroscopic variability among polymers, and (c) to display different physical or chemical properties leading to a variety of potential non-covalent interactions between polymer and analyte. For this study, excellent results were obtained with an array of 13 polymers. This number may be modified depending on the responses elicited by the analytes under consideration. The number of polymers was also intentionally kept to a minimum to expedite data processing and to allow for future optimization. Over 630 different combinations are available [17].
6. For deeper wells, a combination of 504 photoresist and SiO_2 hard mask was used. In this approach, 4 μm oxide films were

deposited by plasma-enhanced chemical vapor deposition (PECVD) and the resulting oxide patterned with conventional 504 photoresist. The exposed oxide areas were etched with buffer oxide etch (BOE). The same RIE conditions were used to etch the oxide.

7. If RIE is not available or possible to form the microarray, one alternative is to deposit microspots of the BP sensors on a microscope slide or quartz coverslip. Simply deposit a small amount of BPs in chloroform. Next, deposit the beads on the slide by scooping the beads out using a microspatula and swiping them on the slide. Chloroform was chosen because it has a low Raman signal, causes beads to swell slightly and stick to the slide, and evaporates quickly, leaving dry BP microspots behind.

8. To create sensor arrays that are tailored to detect a specific disease or biomarker(s) of disease, sensors that have the strongest responses to the most relevant analytes of that disease should be chosen. First, create a library of sensor responses to both individual analytes and analyte mixtures at clinically relevant concentrations (usually ppm or ppb), which can then be cross-referenced with real patient samples. The BPs that show the strongest response (highest θ) and/or most selective response (unique variations in the spectra) should be used to fabricate the array. Successful completion of this iterative process will result in a highly sensitive, disease-specific e-nose or e-tongue sensor array.

References

1. Walt DR, Stitzel SE, Aernecke MJ (2012) Artificial noses. Am Sci 100:38–45. https://doi.org/10.1146/annurev-bioeng-071910-124633
2. Gardner JW, Bartlett PN (1994) A brief history of electronic noses. Sensors Actuators B Chem 18:210–211. https://doi.org/10.1016/0925-4005(94)87085-3
3. Falasconi M, Concina I, Gobbi E et al (2012) Electronic nose for microbiological quality control of food products. Int J Electrochem 2012:1–12. https://doi.org/10.1155/2012/715763
4. Star A, Joshi V, Skarupo S et al (2006) Gas sensor array based on metal-decorated carbon nanotubes. J Phys Chem B 110:21014–21020. https://doi.org/10.1021/jp064371z
5. Fitzgerald J, Fenniri H (2017) Cutting edge methods for non-invasive disease diagnosis using E-tongue and E-nose devices. Biosensors 7. https://doi.org/10.3390/bios7040059
6. Albert KJ, Myrick ML, Brown SB et al (2001) Field-deployable sniffer for 2,4-dinitrotoluene detection. Environ Sci Technol 35:3193–3200. https://doi.org/10.1021/es010829t
7. Fitzgerald JE, Fenniri H (2016) Biomimetic cross-reactive sensor arrays: prospects in biodiagnostics. RSC Adv 6:80468–80484. https://doi.org/10.1039/C6RA16403J
8. Katada S, Hirokawa T, Oka Y et al (2005) Structural basis for a broad but selective ligand spectrum of a mouse olfactory receptor: mapping the odorant-binding site. J Neurosci 25:1806–1815. https://doi.org/10.1523/JNEUROSCI.4723-04.2005
9. Buck LB (2005) Unraveling the sense of smell (Nobel Lecture). Angew Chem Int Ed 44:6128–6140. https://doi.org/10.1002/anie.200501120
10. Meixner H, Lampe U (1996) Metal oxide sensors. Sensors Actuators B Chem 33. https://doi.org/10.1109/SENSOR.1995.717380

11. Freund MS, Lewis NS (1995) A chemically diverse conducting polymer-based "electronic nose". Proc Natl Acad Sci U S A 92:2652–2656. https://doi.org/10.1073/pnas.92.7.2652
12. Grate JW (2000) Acoustic wave microsensor arrays for vapor sensing. Chem Rev 100:2627–2648. https://doi.org/10.1021/cr980094j
13. Liu L, Wang X, Ma Q et al (2016) Multiplex electrochemiluminescence DNA sensor for determination of hepatitis B virus and hepatitis C virus based on multicolor quantum dots and Au nanoparticles. Anal Chim Acta 916:92–101. https://doi.org/10.1016/j.aca.2016.02.024
14. Sisk BC, Lewis NS (2006) Vapor sensing using polymer/carbon black composites in the percolative conduction regime. Langmuir 22:7928–7935. https://doi.org/10.1021/la053287s
15. Lim SH, Kemling JW, Feng L, Suslick KS (2009) A colorimetric sensor array of porous pigments. Analyst 134:2453–2457. https://doi.org/10.1039/b916571a
16. Walt DR (2010) Bead-based optical fiber arrays for artificial olfaction. Curr Opin Chem Biol 14:767–770. https://doi.org/10.1016/j.cbpa.2010.06.181
17. Fenniri H, Chun S, Terreau O, Bravo-Vasquez JP (2008) Preparation and infrared/Raman classification of 630 spectroscopically encoded styrene copolymers. J Comb Chem 10:31–36. https://doi.org/10.1021/cc7001292
18. Fitzgerald JE, Zhu J, Bravo-Vasquez JP, Fenniri H (2016) Cross-reactive, self-encoded polymer film arrays for sensor applications. RSC Adv 6:82616–82624. https://doi.org/10.1039/C6RA13874H
19. Fitzgerald JE, Shokat Fadaee S, Sundaram R, Fenniri H (2019) Plastic-based sensor arrays: towards an affordable diagnosis of Hepatitis C virus. Sensors Actuat B-Chem 288:332–336. https://doi.org/10.1016/j.snb.2019.03.023
20. Arshady R, Ledwith A (1983) Suspension polymerisation and its application to the perparation of polymer supports. React Polym 1:159–174
21. Fenniri H, Ding L, Ribbe AE, Zyrianov Y (2001) Barcoded resins: a new concept for polymer-supported combinatorial library self-deconvolution [18]. J Am Chem Soc 123:8151–8152. https://doi.org/10.1021/ja016375h
22. Fenniri H, Chun S, Ding L et al (2003) Preparation, physical properties, on-bead binding assay and spectroscopic reliability of 25 barcoded polystyrene - poly(ethylene glycol) graft copolymers. J Am Chem Soc 125:10546–10560. https://doi.org/10.1021/ja035665q

Chapter 2

Interferometric Reflectance Imaging Sensor (IRIS) for Molecular Kinetics with a Low-Cost, Disposable Fluidic Cartridge

James W. Needham, Nese Lortlar Ünlü, Celalettin Yurdakul, and M. Selim Ünlü

Abstract

The determination of kinetic information and appropriate binding pairs is fundamental to the proper optimization and selection of ligands used in immunoassays, diagnostics, and therapeutics. However, the ability to estimate such parameters in a multiplexed and inexpensive format remains difficult and modification of the ligand is often necessary. Here, we detail the methods and materials necessary to evaluate hundreds of unlabeled ligands simultaneously using the interferometric reflectance imaging sensor (IRIS). The incorporation of a low-cost fluidic cartridge that integrates on the top of the sensor simplifies reagent handling considerably.

Key words Interferometric reflectance imaging sensor (IRIS), Multiplexed kinetics, Label-free binding, Dengue NS1

1 Introduction

The measurement and understanding of the kinetic data involved in ligand-analyte binding events (e.g., the typical antibody-antigen complex) are foundational to immunoassay development and optimization [1]. A variety of methods exist for the determination of such kinetic information, including surface plasmon resonance (SPR) and biolayer interferometry (BLI), among others [2, 3]. However, most methods are limited in scalability (by cost or technology) with only one or a handful of ligands that may be evaluated at a given time, or the necessity for labeling of the ligand (e.g., biotinylation) in order to perform the assay. Additionally, evaporation may cause some artifacts with data collection using BLI [4]. Here, we describe the methods to use the interferometric reflectance imaging sensor (IRIS) in order to screen hundreds and potentially thousands of ligands simultaneously using an

Jessica E. Fitzgerald and Hicham Fenniri (eds.), *Biomimetic Sensing: Methods and Protocols*, Methods in Molecular Biology, vol. 2027, https://doi.org/10.1007/978-1-4939-9616-2_2,

inexpensive disposable chip and cartridge. We also note that while the example described here is specific to protein-protein interactions, the IRIS instrument also readily detects dynamic nucleic acid binding events [5, 6]. The low cost of both the IRIS and chip gives the ability to characterize the kinetic information for multiple ligands rapidly and affordably. In addition, the low volume of analyte required permits the efficient use of potentially precious reagents and/or the evaluation of real serum specimens.

An example binding assay is shown here that demonstrates kinetic binding of dengue nonstructural protein 1 (NS1) to various monoclonal antibodies. Multiple elution and binding steps are also shown, demonstrating the varying affinities of the monoclonal antibodies to each dengue serotype protein.

2 Materials

All reagents are prepared using highly purified water (resistivity >18 MΩ) unless otherwise stated. Buffers that are used in the IRIS system are brought to room temperature and filtered with a 0.22 μm filter prior to use.

2.1 IRIS System

The basic layout of the IRIS system is shown in Fig. 1 and a photograph of the complete system is shown in Fig. 2. Fundamentally, a light source (light-emitting diode) illuminates the chip surface. The light reflects from both the silicon and oxide layers and generates an interference pattern that can be monitored as an intensity change with respect to oxide thickness at the camera. The

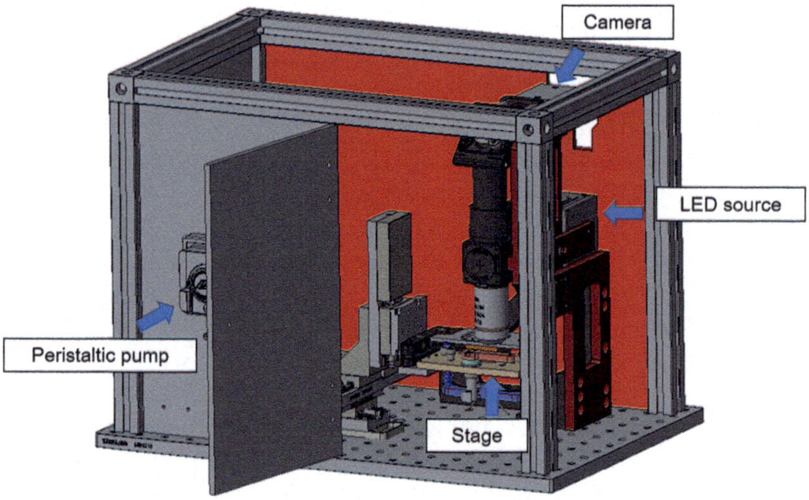

Fig. 1 A representation of the IRIS system. Locations of the LED source, camera, stage, and peristaltic pump are indicated

Fig. 2 The physical benchtop setup of the IRIS system

reflectance formula for a single-layer thin film at the far field with a normal incidence angle can be estimated by [7]

$$R = |r|^2 = \frac{r_1^2 + r_2^2 + 2r_1 r_2 \cos(2\varphi)}{1 + r_1^2 r_2^2 + 2r_1 r_2 \cos(2\varphi)} \tag{1}$$

where

$$\varphi = \frac{2\pi d}{\lambda} n_{\mathrm{SiO_2}}$$

and r_1 and r_2 are Fresnel reflection coefficients of *water/SiO*$_2$ and *SiO*$_2$*/Si* interfaces, respectively; d is the thickness of the silicon oxide; and n_{SiO_2} is the wavelength-dependent refractive index of the oxide layer.

The key components of the IRIS system include the following:

1. LEDs: A spatiotemporally low coherent chip-on-board (COB) four-color LED is mounted onto an integrating sphere (IS236A-4, Thorlabs) which provides an approximately uniform output beam by multiple reflections of the incoming light. The integrating sphere enables the use of critical illumination techniques without presenting unwanted LED structures onto the camera sensor. In this manner, the illumination arm becomes compact and solid. The COB-LED consists of four different dyes to excite at four colors: red (dominant wavelength, $\lambda_D = 630$ nm), green ($\lambda_D = 525$ nm), blue ($\lambda_D = 460$ nm), and yellow ($\lambda_D = 595$ nm) (LZ4-00MA00, LED Engin). At the initiation of an experiment, four-color images of the chip may be acquired for spectral curve fitting of the oxide layer thickness. Once the oxide layer is

determined, the blue wavelength may be used for the rest of the experiment.

2. Turn mirror: A silver-coated, 45° elliptical mirror (PFE10-P01, Thorlabs) is placed on a kinematic-elliptical mount to rotate the collection path by 90°. The elliptical mirror has a clear aperture at 45° and is less prone to clip the collected light beam.
3. XYZ stage: The stage may be either manually or computer controlled to bring the chip into the proper position and into an appropriate focus. Once the correct position and focus are determined for a given experiment, the positions remain unchanged.
4. Stage: The stage holds the chip in place, provides the contact points through which tubing and O-rings come into contact with the chip, and applies pressure to the antireflective (AR)-coated glass slide and gasket to create the fluidic cartridge. A representation of the stage is shown in Fig. 3 [8].
5. Camera: A 5 MP (2448 × 2048) global shutter CMOS camera is used. The pixel pitch of the camera is 3.45 μm and the camera sensor size is 8.445 mm × 7.065 mm. The total field of view (FOV) is limited by the camera sensor and the magnification provided by the objective. An approximate 4.23 mm × 3.5 mm FOV, for instance, can be obtained by using a 2× microscope objective, whereas a 5× objective provides only a 1.7 mm × 1.4 mm FOV. Additionally, cameras with a larger sensor size can be employed to increase the total FOV.
6. Objective: A 2×/0.06 NA or 5×/0.1 NA objective (Nikon) may be used as desired (*see* **Note 1**).

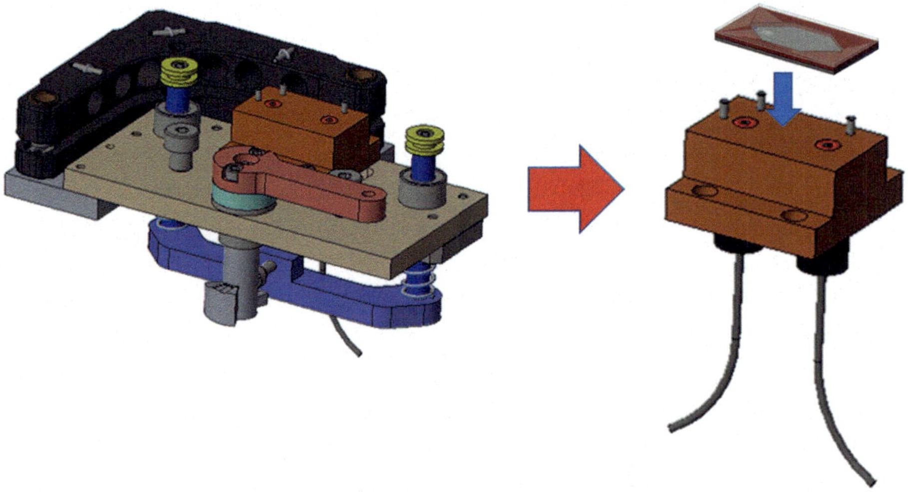

Fig. 3 The stage and integration of the IRIS chip onto the stage surface. Note the O-rings and tubing that are responsible for the delivery and removal of all reagents across the chip surface

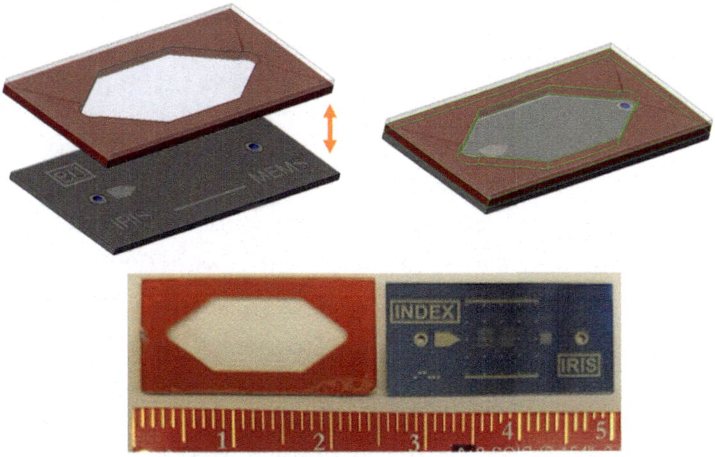

Fig. 4 The IRIS chip and antireflective (AR) slide with gasket. An example spotted array (still "wet") is visible on the IRIS chip surface

2.2 Chip and Fluidic Cartridge

The IRIS chip forms a union with an AR-coated glass slide that is adhered to a thin silicone gasket. Due to the presence of through-silicon-via (TSV) holes, the chip and glass slide form a small, sealed chamber through which desired buffers may be pumped.

The chip and cartridge are detailed below and in Fig. 4:

1. Silicon chip (25.2 mm × 12.5 mm) with a 110 nm thermally grown oxide layer.
2. Through-silicon-via (TSV) holes placed appropriately to permit fluid flow over the chip surface.
3. An appropriately sized AR glass slide is attached with a gasket. Alternatively, an optically transparent coverslip with pressure-sensitive adhesive (PSA) may be used (*see* **Note 2**).

2.3 Spotter

1. BioRad BioOdyssey Calligrapher: Any spotter capable of delivering ~100 μm sized spots at 200–300 μm pitch is sufficient (*see* **Notes 1**, **3**, and **4**).

2.4 Fluidics

1. Peristaltic pump: A variable-speed peristaltic pump and appropriate connectors are necessary for the delivery of all reagents over the chip surface (*see* **Note 5**).
2. De-bubbler: An inline de-bubbler (Elveflow) is used to remove bubbles before solutions reach the chip surface (*see* **Note 6**).
3. Appropriately sized tubing and connectors.

2.5 Buffers

1. 3% (v/v) Epoxysilane: 3-(glycidoxypropyl)trimethoxysilane ("epoxysilane") is diluted into 100% isopropanol immediately

prior to use. The epoxysilane is moisture sensitive and care should be taken when handing. A needle and syringe should be used to remove the required volume. For example, 300 μL of epoxysilane is added directly to 9.7 mL of 100% isopropanol into a polypropylene test tube. The tube is briefly mixed by inversion prior to submerging the chip surface.

2. 100% Isopropanol: Molecular biology grade.
3. 50% (v/v) Glycerol: 100% Glycerol is diluted into highly purified water.
4. 10× PBS, pH 6.9: 100 mM Sodium phosphate dibasic with 1.5 M sodium chloride.
5. 1× PBS, pH 7.4: 1× PBS is prepared by diluting 10× PBS into highly purified water.
6. 10% (v/v) Tween-20: 100% Polysorbate 20 ("Tween-20") is diluted into highly purified water.
7. Assay buffer: 1% (w/v) BSA in 1× PBS, 0.1% (v/v) Tween-20 (1% BSA in PBST): Bovine serum albumin (BSA) is diluted into 1× PBS. 10% Tween-20 is added to the solution to a final concentration of 0.1% (v/v).
8. Wash buffer: 1× PBS, 0.1% Tween-20 (1× PBST): 10% Tween-20 is added to 1× PBS to a final concentration of 0.1% (v/v).
9. Spotting buffer: 1× PBS containing 1% (v/v) glycerol. 50% Glycerol is added to the volume of antibody (or other protein) used for spotting such that the final concentration is 1%.
10. Elution buffer: 0.1 M Glycine, pH 2.5.

2.6 Antibodies and Antigens

The actual conditions and concentrations used for the analytes of interest will, as a matter of course, vary on a case-by-case basis. The conditions described below provide an example method for spotting antibodies onto an activated chip surface and flowing analyte (antigens) over the chip surface to visualize ligand-analyte binding events. The specific example described here uses dengue NS1 antigen as the analyte and a series of anti-NS1 monoclonal antibodies that are spotted onto the chip.

1. Stock antigens: Antigens need to be at a sufficiently high concentration in order to prepare the ready-to-use antigen. The antigen used in this example study is dengue-1 NS1 antigen, dengue-2 NS1 antigen, dengue-3 NS1 antigen, and dengue-4 NS1 antigen (Native Antigen Company).
2. Stock antibodies: Antibodies should be at a sufficiently high concentration for spotting onto the activated chip surface. Concentration ranges should be 0.5–2 mg/mL. The antibodies used in this example were kindly provided by InBios

International, Inc. Spots of BSA and an irrelevant antibody were included as negative controls.

3. Spotting antibody solution: The reagents that are to be spotted onto the chip should be between 0.5 and 2 mg/mL in 1× PBS containing 1% (v/v) glycerol. The appropriate spotting conditions may vary depending upon the spotter used.
4. Preparation of ready-to-use (RTU) antigen: The antigen (in this case, dengue NS1) should be diluted into 1% BSA in PBST to the appropriate concentration needed for the assay and then filtered (*see* **Notes 7** and **8**). Typical concentrations may range from 10 ng/mL to 100 μg/mL. The concentrations used in this example study are at 10 μg/mL (~220 nM) in 1% BSA in PBST.

2.7 Software

1. *Image acquisition*: Customized ImageJ (Micromanager) software [9] is used to control the camera and capture images. Images are taken every 6–12 s at a single wavelength. Multiple wavelengths ($n = 4$) may also be used in order to estimate the thickness (nm) of the oxide surface and the subsequent binding events that occur.
2. *Data analysis*: ImageJ is used to acquire all images and determine the mean pixel value in a given spot. The mean value in the region of interest is determined and a corresponding mean value in a nearby "background" region is also calculated. The difference between the mean value within the spot and the background region constitutes the signal that is observed. If multiple wavelengths are used, the nm thickness increase may be determined. Otherwise, the pixel value (AU) is considered approximately proportional to the thickness increase observed. Signals are passed through a simple low-pass Butterworth filter prior to analysis (to remove high-frequency noise). On and off rates (k_{on} and k_{off}) may then be estimated using custom scripts written in Python or other available software, as desired.

3 Methods

3.1 Chip Functionalization

Several chips may be functionalized at a time and subsequently stored in desiccating conditions (e.g., in the presence of desiccant or under vacuum) for several weeks prior to use. A variety of methods have been used to functionalize silicon oxide surfaces. We have successfully functionalized chip surfaces using both a polymeric coating (MCP-2, Lucidant Polymers) [10] and epoxysilane. The method described here focuses on a simple epoxysilanization method to functionalize multiple chips. Use dedicated glassware or disposable petri dishes to complete the following (all steps at room temperature):

1. Submerge chip(s) in 1 N NaOH for 1 h, gently moving on a rocker platform.
2. Briefly rinse chip(s) in a large volume (~1 L) of dIH_2O to ensure that all salts have been removed.
3. Dry the chip(s) thoroughly under nitrogen and place into a clean petri dish.
4. Immediately prepare a *fresh* epoxysilane solution, 3% (v/v) in 100% isopropanol.
5. Submerge the dried chip(s) in the epoxysilane solution for 3 min, gently swirling or rocking.
6. Rinse the chip(s) in 100% isopropanol for 5 min.
7. Dry thoroughly under nitrogen (or other inert gases).
8. Place in a vacuum desiccator for 1 h.
9. Store the chip(s) at room temperature in the presence of desiccant (or in a vacuum desiccator).

3.2 Spotting

Chips are contact spotted using a BioRad BioOdyssey Calligrapher with a 100 μm capillary pin. Antibodies and controls spotted onto a chip surface typically range from 0.5 to 2 mg/mL, containing 1% (v/v) glycerol to aid spot morphology. After spotting, leave chips overnight at room temperature to permit the antibodies to fully bind to the chip (*see* **Note 3**).

3.3 Washing and Blocking

After spotting, chips are blocked using a 1% BSA in PBST buffer.

1. First hold chip at an angle and pour 1% BSA in PBST over the surface of the chip.
2. Rinse the chip three times in 1× PBST (*ca.* 30 s with each rinse, submerged in a petri dish containing 1× PBST).
3. Place chip in a small dish and completely submerge in 1% BSA in PBST for 5 min to block.
4. After this blocking step, rinse the chip in 1× PBST, dry gently under nitrogen, and load into the IRIS to run the assay.

3.4 Loading the Chip into the IRIS

The microfluidic cartridge is created by the union of the chip (with the existing holes predesigned in the surface) with the boundary layer of sealed, optically transparent layer placed on top. Two options are used for creating this microfluidic cartridge. One option places a custom-designed (GraceBio) pressure-sensitive adhesive cover on top of the chip. Adhesive is present only around the edges of the chip surface, forming a barrier such that fluid may pass from one port to the other. A second option uses an antireflective (AR)-coated window with a custom-cut gasket attached to the surface. With either option:

1. Apply the top portion of the cartridge to the chip surface and ensure that a good seal is maintained around the chip edge (*see* **Note 2**).
2. Carefully load the chip onto the stage.
3. Make sure that the holes in the chip align with the O-rings present on the stage surface.
4. Lower the upper portion of the stage to lock the chip into place.

3.5 Running a Typical Binding Assay

Once the chip is loaded, the binding assay can proceed as desired. A typical binding assay is described below, using the dengue NS1 antigen as an example case. Dengue virus has four distinct serotypes; infection with one serotype does not necessarily provide immunity to subsequent infections with another serotype [11]. Nonstructural protein 1 (NS1) is an early marker for dengue infection, detectable within the first few days after clinical symptoms [12]. However, the antigenic sequences for nonstructural protein 1 (NS1) also vary by serotype and determining high-affinity, specific, and/or cross-reactive antibodies may be of significant diagnostic utility [13]. In this example, we look at real-time binding events for each serotype with a panel of antibodies.

1. Start acquiring the images using ImageJ/Micromanager software. Time points of acquisition can be determined in the software but acquiring an image every 6–12 s has been adequate.
2. Place the collected images into an image stack in ImageJ.
3. Switch between buffers manually by simply placing the inlet tubing into the desired solution. Small bubbles that enter into the tubing during this switch are eliminated by the de-bubbler prior to the solution reaching the chip surface.

   A typical assay process consists of the following:

1. Run 1% BSA in 1× PBST over the chip for 10–20 min (*see* **Note 9**). Typical flow rates have been used at 100–200 μL/min.
2. Run NS1 antigen (10 μg/mL diluted into the 1% BSA in 1× PBST buffer—*see* **Note 2**) over the chip for 20 min.
3. Run 1% BSA in 1× PBST over the chip for 40 min.
4. Elute the NS1 antigen using 0.1 M glycine, pH 2.5, for 5 min.
5. Run 1% BSA in 1× PBST over the chip for 10 min to equilibrate.

At this point, it may be desirable to test the same set of antibodies using a different analyte. Alternatively, sandwiching antibodies may be added after **step 3** in order to determine possible binding pairs. This typical procedure may be repeated several times, but the activity of each antibody or ligand bound to the surface may be impacted by the elution step.

3.6 Typical Result and Analysis

1. Obtain binding curves for each of the spots printed on the chip.
2. Using ImageJ software, select a region of interest for each spot and a nearby corresponding background region of interest is selected as a reference for that same spot.
3. Calculate the mean value within the spot and the mean value in the region of interest using the software for each image in the stack.
4. The dynamic binding data points are then simply the difference between the mean value within the spot and the mean value from the reference region of interest [14].

If desired, it is possible to create a series of images that shows the binding events in a movie format (.avi). In this case, the first image that was acquired is effectively subtracted from all of the subsequent images in the stack (*see* **Note 10**). The subsequent binding events are then more readily visualized due to the accumulation of mass at the individual spot.

Experiments were performed at InBios International, Inc. (Seattle, WA) and typical results for the acquired images and binding curves as shown in Figs. 5, 6, and 7. A supplementary video file (S1) is also provided that shows the real-time binding of the analytes to the various ligands.

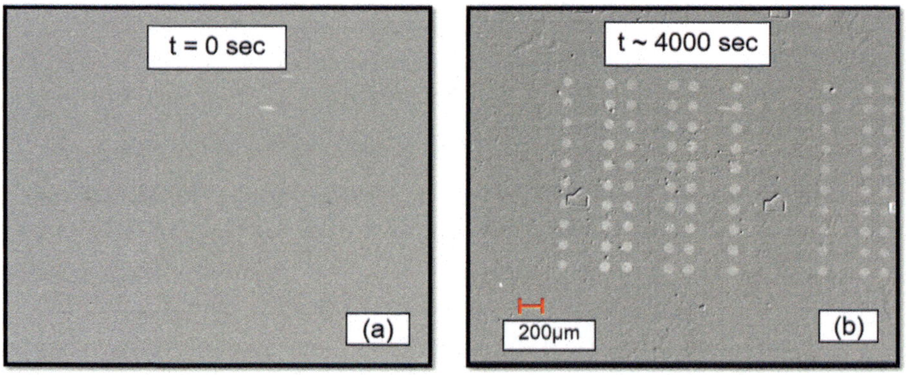

Fig. 5 Example time points (slices) of the binding assay. The start of the assay is represented in (**a**) where no binding has occurred and (**b**) shows a time point after ~4000 s. The supplementary video file (S1) shows a compressed time course for all of the binding and elution events from this example system. Reagents were spotted in vertical replicates of $n = 10$ each

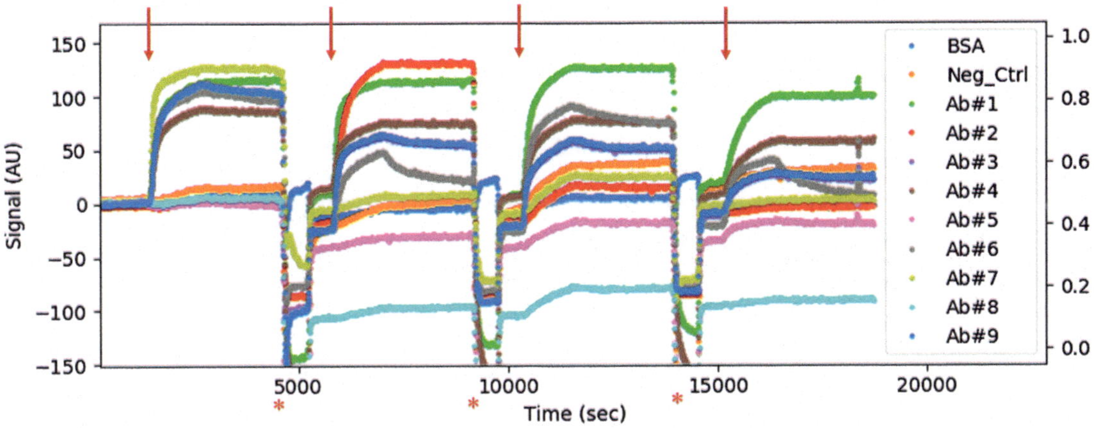

Fig. 6 The dynamic binding curves collected with the example system tested. Binding curves were collected for dengue-1, -2, -3, and -4 NS1 protein binding. Each curve represents the average of five individual spots. Red arrows indicate points at which dengue-1, -2, -3, and -4 NS1 proteins were sequentially injected into the fluidic cartridge, respectively. A red (∗) indicates an elution and regeneration step

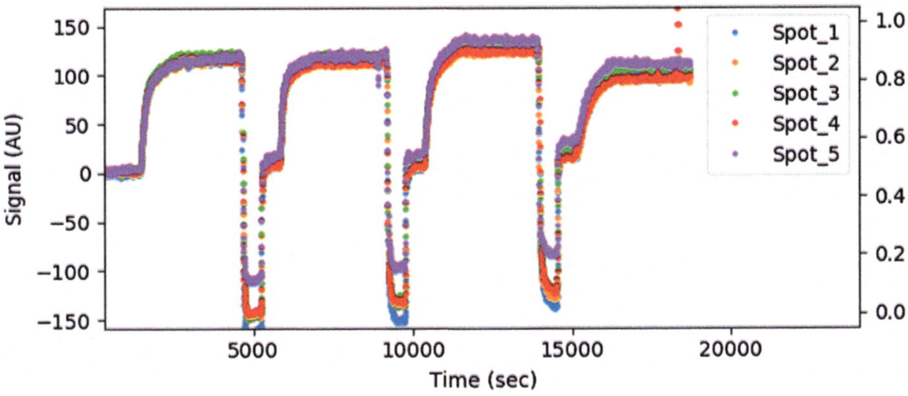

Fig. 7 The data collected for individual spots with Ab#1. See the indicators in Fig. 6 regarding injection and elution steps

4 Notes

1. **Noise control**: The noise observed in the assay may be variedly controlled to a large degree by the user. Simply changing the objective used from 2× to 5× significantly increases the number of "pixels per spot" and the subsequent noise floor may be reduced. Alternatively, a larger spot size may be used to print the ligands of interest and thus further reduce the desired noise floor.

2. **Applying the pressure-sensitive adhesive (PSA) coverslip**: There are two protective layers that are part of the PSA coverslip. One layer protects the transparent upper layer while the second layer protects the adhesive portion until it is ready to use. When using the PSA coverslip, use precision tweezers to remove the clear adhesive layer. Carefully line up the cover over the IRIS chip and gently apply pressure to adhere the cover to the chip. It is helpful to actually pick up the chip by hand and align the adhesive cover to the chip with gloved fingers. After the cover is on, place the chip on a flat surface, use a clean small polypropylene test tube, and "roll" the tube several times over the cover to apply pressure evenly across the chip surface. After the adhesive is completely attached, start to peel the protective upper layer on the coverslip. Ensure that nothing touches the transparent surface once the protective layer is removed.
3. **Spot quality and morphology**: Spotting should be sufficiently uniform to ensure reliable results. It is useful and simple to use the IRIS to check for spot consistency (i.e., any "missing" spots) and morphology prior to running the assay. Checking for spot consistency both prior to washing and blocking and subsequently post-washing and -blocking (and prior to running any buffer) may be useful to verify chip quality prior to running the assay.
4. **Spotting with the calligrapher**: It was found useful to include ~0.01% Tween-20 in the wash buffer basin when using the BioRad BioOdyssey Calligrapher in order to ensure consistent spotting with the 100 μm pin.
5. **Peristaltic pump**: Binding assays have been successfully performed at rates of ~100–200 μL/min. Flowing buffer over the chip too quickly may cause fluctuations with the soft PSA coverslip (but not with the glass AR slide). Flowing buffer too slowly will not change the reagent volume sufficiently often and potentially lead to diffusion limitations in the assay.
6. **Pre-filling fluidic chamber**: In order to reduce/eliminate bubble formation, it is simple to preload the fluidic chamber with buffer by simply pipetting the buffer into the chamber until it is completely filled. Hold the chamber at an angle such that any air in the chamber is completely removed while the chamber is filled. The chamber can then be loaded on a (primed) stage holder and bubbles in the chamber may thus be avoided.
7. **Matching buffer refractive index**: It is necessary to use a running buffer that is the primary diluent of the analyte of interest. This will prevent changes in refractive index impacting the measured result when evaluating single-wavelength intensity values only.

8. **Filtering buffers**: It is necessary to properly filter reagents (0.22 μm syringe filter) prior to flowing over the IRIS chip; if this precaution is not followed, extraneous particles may flow past the viewing window during image acquisition and cause unnecessary noise.
9. **Equilibrating IRIS chip**: Before initiating the assay, ensure that the chip is equilibrated sufficiently. That is, run buffer over the chip for some time while collecting data. Select one or two control spots and evaluate the signal at the spot to make sure that a steady state has been reached. This typically is observed within 20–30 min after loading the chip into the IRIS stage.
10. **Generating a "difference" movie**: To visualize binding events, the first image in the collected image stack may be used as a "reference" image. An appropriate constant value may be subtracted from this "reference" image in order to avoid bottoming out subsequent subtraction steps. To avoid large file sizes, a copy of the image stack can be resized to make handling the data more manageable. The "reference" image is then subtracted from each subsequent image in the image stack (using ImageJ) and the file is saved as a .avi file.

References

1. Wild D (ed) (2005) The immunoassay handbook. Elsevier Ltd., Oxford
2. Boozer C, Kim G, Cong S, Guan H, Londergan T (2006) Looking towards label-free biomolecular interaction. Curr Opin Biotechnol 17:400–405
3. Yang D, Singh A, Wu H, Kroe-Barrett R (2017) Determination of high-affinity antibody-antigen binding kinetics using four biosensor platforms. J Vis Exp. https://doi.org/10.3791/55659
4. Naman S, Duncan T (2014) Bio-layer interferometry for measuring kinetics of protein-protein interactions and allosteric ligand effects. J Vis Exp. https://doi.org/10.3791/51383
5. Özkumur E, Ahn S, Yalçin A, Lopez C, Çevik E, Irani R, DeLisi C, Chiari M, Ünlü MS (2010) Label-free microarray imaging for direct detection of DNA hybridization and single-nucleotide mismatches. Biosens Bioelectron 25:1789–1795
6. Zhang X, Daaboul GG, Spuhler PS, Freedman DS, Yurt A, Ahn S, Avci O, Ünlü MS (2014) Nanoscale characterization of DNA conformation using dual-color fluorescence axial localization and label-free biosensing. Analyst 139:6440–6449
7. Özkumur E, Needham J, Bergstein D, Gonzalez R, Cabodi M, Gershoni J, Goldberg B, Ünlü MS (2008) Label-free and dynamic detection of biomolecular interactions for high-throughput microarray applications. Proc Natl Acad Sci U S A 105(23):7988–7992
8. Trueb J (2018) Enabling and understanding nanoparticle surface binding assays with interferometric imaging. Dissertation, Boston University
9. Abramoff M, Magalhaes P, Ram S (2004) Image processing with ImageJ. Biophoton Int 11(7):36–42
10. Seymour E, Daaboul G, Zhang X, Scherr S, ÜnlÜ N, Connor J, ÜnlÜ MS (2015) DNA-Directed antibody immobilization for enhanced detection of single viral pathogens. Anal Chem 87:10505–10512
11. Murrell S, Wu SC, Butler M (2011) Review of dengue virus and development of a vaccine. Biotechnol Adv 29(2):239–247
12. Lima MRQ, Nogueira RMR, Schatzmayr HG, dos Santos FB (2010) Comparison of three commercially available dengue NS1 antigen capture assays for acute diagnosis of dengue in Brazil. PLoS Negl Trop Dis 4(7):e378. https://doi.org/10.1371/journal.pntd.0000738

13. Röltgen K, Rose N, Ruggieri A, Warryn L, Scherr N, Pinho-Nascimento CA, Tamborrini M, Jaenisch T, Pluschke G (2018) Development of dengue virus serotype–specific NS1 capture assays for the rapid and highly sensitive identification of the infecting serotype in human sera. J Immunol 200 (11):3857–3866

14. Vedula R, Daaboul G, Reddington A, Özkumur E, Bergstein D, Ünlü MS (2010) Self-referencing substrates for optical interferometric biosensors. J Mod Opt 57 (16):1564–1569

Chapter 3

An Olfactory Sensor Array for Predicting Chemical Odor Characteristics from Mass Spectra with Deep Learning

Yuji Nozaki and Takamichi Nakamoto

Abstract

Machine learning techniques are useful for applications such as electronic nose (e-nose) systems to classify or identify the target odor. In recent years, deep learning is regarded as one of the most powerful machine learning methods. It enables researchers to extract useful features automatically from high-dimensional raw data and has been widely applied to computer vision, speech recognition, and natural language processing, though little has been reported in the field of olfaction. In this chapter, we describe the procedure to build a deep neural network to predict odor characteristics of chemicals from their mass spectra.

Key words Deep learning, Predictive model, Odor character, Mass spectrum, Dimensionality reduction

1 Introduction

Predictive modeling is a technique which uses mathematical methods to predict outputs (dependent variables, often denoted by y) from given inputs (explanatory variables, often denoted by x). Predictive models vary with application; that is, it is necessary to carefully choose an appropriate model based on the relationship between two or more variables. When the relationship between a dependent variable and an explanatory variable is linear, linear approaches are used for modeling. A linear regression, for example, the height and weight of college students, can fit a simple straight line to the data by the least squares method. When there are more than two explanatory variables, the model is called a multiple linear regression. When the relationship between dependent variable and explanatory variable cannot be expressed by linear function but requires a nonlinear function, a nonlinear approach should be taken. Logistic regression is a popular nonlinear approach in which the standard logistic function is used to relate a binary-dependent variable and an explanatory variable, instead of linear functions.

Jessica E. Fitzgerald and Hicham Fenniri (eds.), *Biomimetic Sensing: Methods and Protocols*, Methods in Molecular Biology, vol. 2027, https://doi.org/10.1007/978-1-4939-9616-2_3,

Neural networks (*see* **Note 1**) and deep learning are currently believed to be one of the best solutions to many problems in image recognition, speech recognition, and natural language processing [1]. Neural networks are also known as very powerful predictive modeling methods [2], as they can be fitted to highly complex, multivariate, nonlinear problems [3]. Neural networks are inspired by the communication mechanisms of the nerve systems of animals, and its nodes are analogous to the neurons of each of those nerve systems. Though artificial neural networks are different from actual biological nerve systems in several aspects, it has shown promising results in a variety of cognition tasks such as image and speech recognition. Since biological olfactory systems are essentially nonlinear [4], neural networks are more appropriate for computational models of olfaction compared to the other modeling techniques we introduced earlier in this chapter. In this chapter, we describe protocols to build a deep predictive neural network based on our previous research [5]. That includes (1) data preprocessing, (2) feature extraction, and (3) modeling protocols.

2 Data Preprocessing

Figure 1 shows the conceptual diagram of odor impression prediction from a mass spectrum. A dataset for target signals (values to be predicted) and a dataset for input signals should be separately prepared. We use the results of sensory evaluation test conducted by A. Dravnieks et al. (hereafter referred to as "sensory data" or "Y") [6] for the target signal, and its mass spectrum for input signals (hereafter referred to as "mass spectra data" or "X") [7]. 121 chemicals are used in the following example. For the details of two datasets, *see* **Notes 2** and **3**.

1. Transform the raw data into vector representation. Mass spectra correspond to a chemical's physicochemical properties, representing structural information of molecules, and are given as a plot of intensity versus mass-to-charge ratio (m/z), as shown in Fig. 2a.
2. Merge the vectors to form matrix. In the matrix, each row corresponds to the vector representation of a sample and each column indicates m/z. There are n chemicals' mass spectra, and the length of vectors is d, thereby forming the dataset matrix $n \times d$ (Fig. 2b).
3. Normalize (*see* **Note 4**) the dataset by dividing by the maximum value in the dataset to scale values between 0 and 1 (Fig. 2c).
4. Remove unnecessary portions from the original data. A vector representation of mass spectrum typically is of high

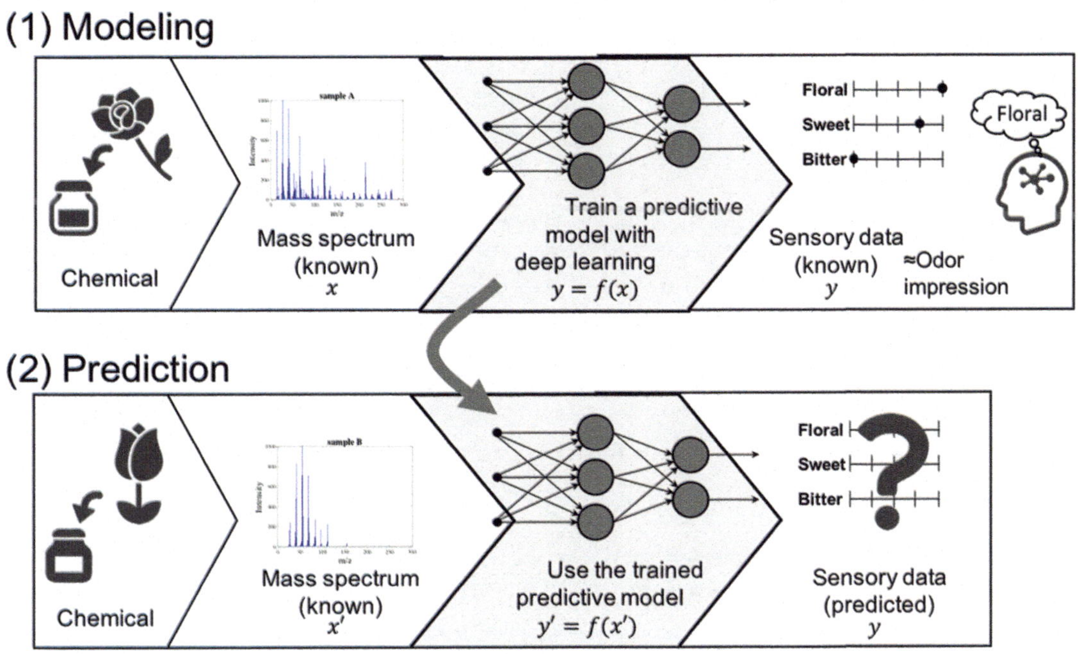

Fig. 1 Conceptual diagram of odor impression prediction

dimensions, i.e., up to more than 1000. We extracted intensities between 51 and 262 m/z, since these intensities are thought to be more representative of the odor character than the other intensities (*see* **Note 5**). After this step, we obtain a dataset matrix of 121 × 212 (Fig. 2d).

5. Obtain the matrix representation of the sensory data by applying the same step. 146 verbal descriptors are originally used in the sensory evaluation. As two descriptors are omitted at **step 4** (*see* **Note 6**), we obtained a 121 × 144 matrix as the sensory data for target signals (Fig. 3).
6. Shuffle the rows in mass spectra dataset, and sort the samples in the sensory dataset so that the samples appear in the same order in both datasets.

3 Methods

In this section, we introduce the procedure to build a predictive model with deep structure. Three neural networks are used to build one predictive model for odor characteristics. Two are deep autoencoders (*see* **Note 7**), which are used for dimensionality reduction (*see* **Note 8**) of each dataset. The other neural network is used to relate feature vectors, which are obtained as a result of dimensionality reduction. A sigmoid function is used for the activation functions (*see* **Note 9**) of the neural network. In the trainings, the back-

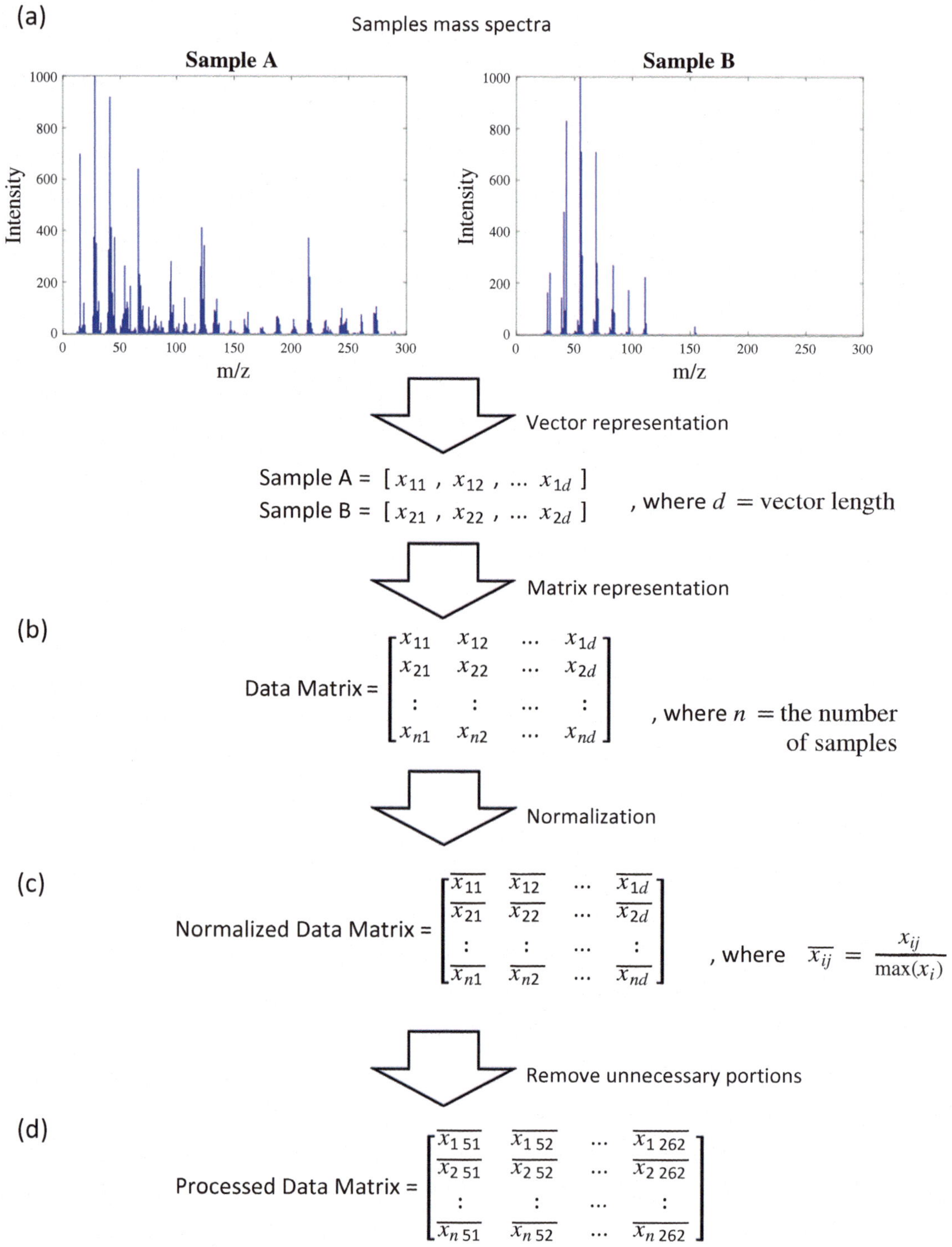

Fig. 2 An example of data preprocessing on mass spectra data. (**a**) Vector representations of samples. (**b**) Matrix representation of the dataset. Each row corresponds to the vector representation of a sample and each column indicates *m*/*z*. (**c**) A normalized data matrix. (**d**) A processed data matrix

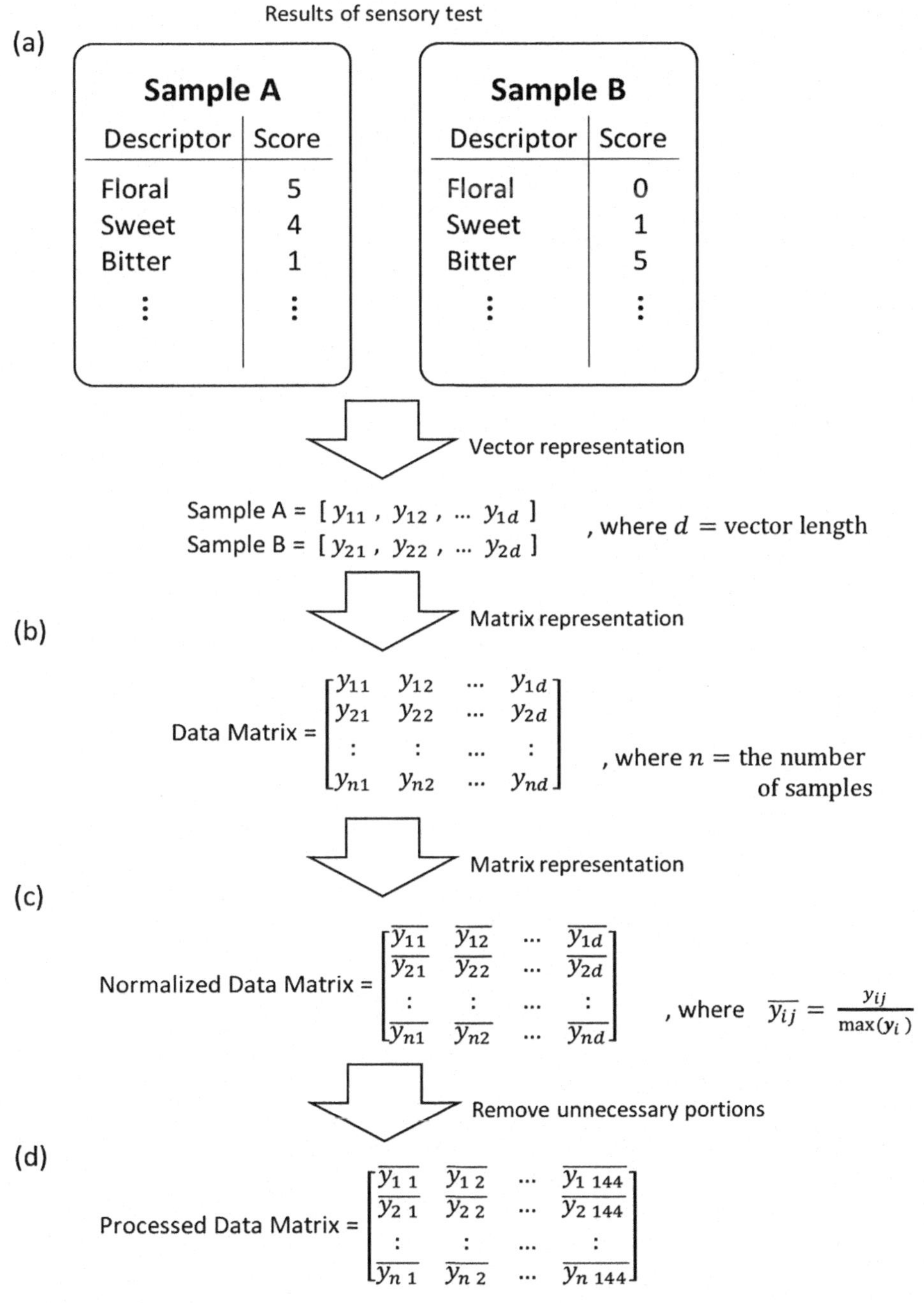

Fig. 3 An example of data preprocessing on sensory data. (**a**) Vector representations of samples. (**b**) Matrix representation of the dataset. Each row corresponds to the vector representation of a sample and each column indicates *m*/*z*. (**c**) A normalized data matrix. (**d**) A processed data matrix

propagation algorithm with the delta rule is used with the stochastic gradient descent algorithm. For details of the updating rule and major techniques we used, such as momentum and regularization, *see* **Note 10**.

3.1 Dimensionality Reduction

To obtain feature representation of mass spectra and sensory data, two deep autoencoders are used. Each autoencoder for dimensionality reduction consists of an input layer, three hidden layers, and an output layer with the same number of neurons in the input layer. Since fewer neurons are set in the middle hidden layer, a low-dimensional representation of an input vector can be obtained.

The deep autoencoders are trained with a layer-wise approach. In this approach, two shallow (three-layer) autoencoders are separately trained and then combined to form a deep (five-layer) autoencoder (*see* **Note 11**).

1. Split the mass spectra dataset into three parts: (1) training ($X_{training}$), (2) validation ($X_{validation}$), and (3) testing ($X_{testing}$), as shown in Fig. 4a (*see* **Note 12**).
2. Apply the same step to sensory data to obtain $Y_{training}$, $Y_{validation}$, and $Y_{testing}$.
3. Build a three-layer autoencoder for dimensionality reduction of mass spectra data (Fig. 5a). The input and the output layers have 212 inputs that correspond to the m/z, and the middle layer has K_M neurons corresponding to the feature vector. Thus, through the following process, original 212-dimensional dataset is compressed into primary feature vectors with K_M dimensions.
4. Train the three-layer autoencoder. For the training, validation, and testing data, $X_{training}$, $X_{validation}$, and $X_{testing}$ are used, respectively.
5. Also initialize hyperparameters, namely the learning rate η, the constant coefficient of the momentum term α, and the number of neurons in layers. Initialize the weights and the biases with a Gaussian distribution with the mean of 0.5 and the standard deviation of 0.1 (*see* **Note 13**).
6. Choose one sample randomly from $X_{training}$ and calculate the output of the neural network with the given weights and biases.
7. Calculate the reconstruct error between the output of the neural network for the training set and the target signal. The reconstruct error is given by the same error function introduced in **Note 10**.
8. Update the weights and the biases with the back-propagation algorithm.
9. Choose one sample randomly from $X_{validation}$ and calculate the output of the neural network with the updated weights and biases.
10. Calculate the reconstruct error between the output of the neural network for the validation set and the target signal.

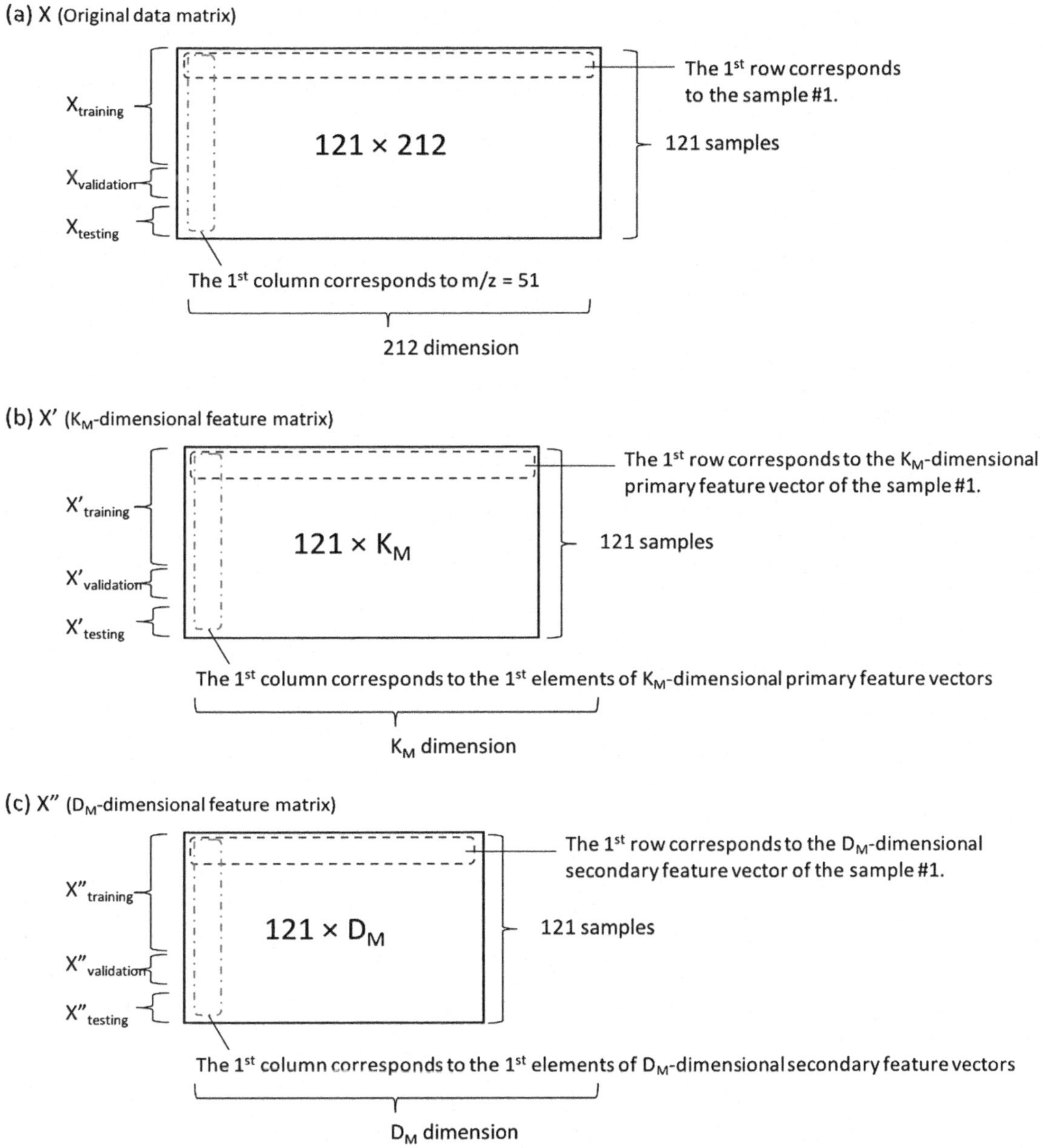

Fig. 4 An example of data partitioning. (**a**) Original matrix of X. (**b**) K_M-dimensional matrix. (**c**) D_M-dimensional matrix

11. Repeat **steps 6–10** until the reconstruct error on $X_{validation}$ converges.
12. Optimize hyperparameters to obtain less reconstruct error for $X_{training}$. Optimization is done by repeating **steps 5–11** after one of the hyperparameters is changed. To obtain better values for hyperparameters, plot errors with respect to the hyperparameter of interest. And then choose the "elbow" point (Fig. 6).

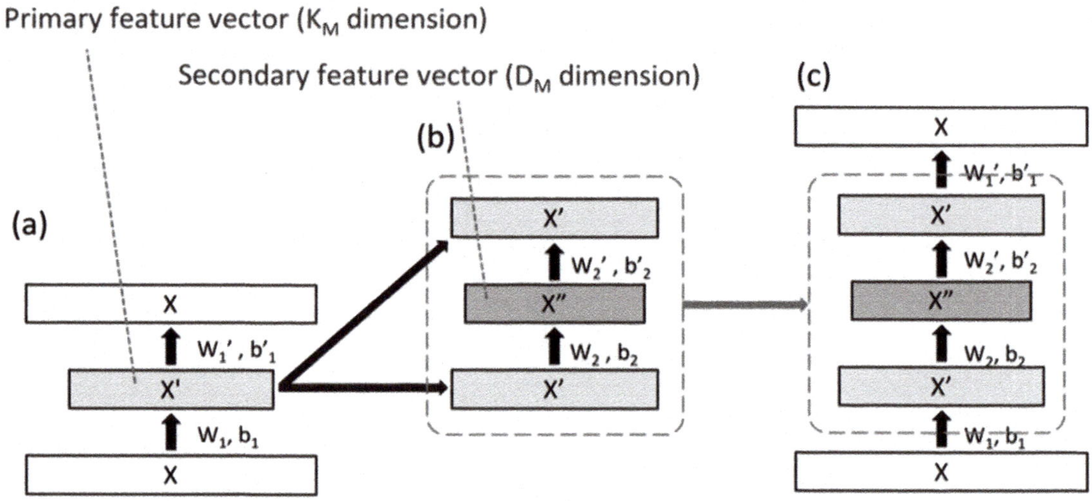

Dimensionality		
212-K_M-212	K_M-D_M-K_M	212-K_M-D_M-K_M-212

Fig. 5 A diagram of autoencoders. (**a**) A diagram of a three-layer autoencoder compressing the original dataset into K_M dimension. (**b**) A diagram of a three-layer autoencoder compressing the K_M-dimensional dataset into D_M dimension. (**c**) A diagram of a five-layer autoencoder compressing the original dataset into D_M dimension

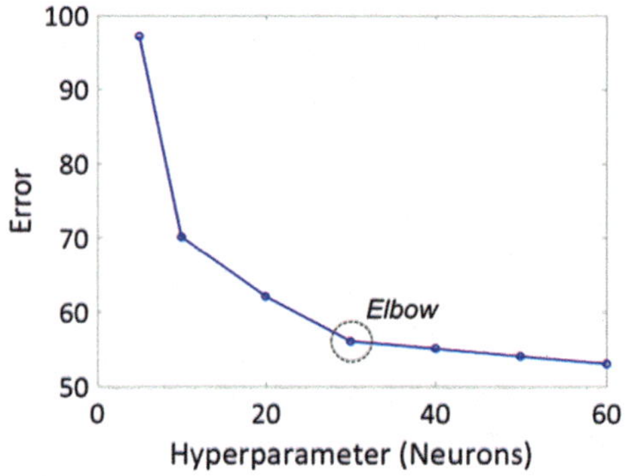

Fig. 6 Reconstruction error as a function of the hyperparameter (number of neurons in a hidden layer)

The elbow point usually represents where the model starts to have only a small change in errors by increasing neurons, which most likely will not influence the accuracy of the model.

13. Save the weights and biases of the neural network as W_1, W_1', b_1, and b_1', respectively. Also save the other hyperparameter values.

14. Obtain K_M-dimensional representations (primary feature vectors) on $X_{training}$, $X_{validation}$, and $X_{testing}$ by transforming raw datasets with saved W_1 and b_1 (hereafter referred to as $X'_{training}$, $X'_{validation}$, and $X'_{testing}$) (shown in Fig. 4b).
15. Build and train another three-layer autoencoder (shown in Fig. 4b). The input and the output layers have K_M neurons, and the middle layer has D_M neurons. For the training, validation, and testing data, $X'_{training}$, $X'_{validation}$, and $X'_{testing}$ are used instead of the original datasets. By following the same step, obtain weights and biases (W_2, W_2', b_2, and b_2'). This now compresses the K_M-dimensional dataset into secondary feature vectors of D_M dimensions (shown in Fig. 4c). A five-layer autoencoder is formed by combining the two autoencoders (shown in Fig. 5c).
16. Initialize the weights and the biases of the five-layer autoencoder with the saved weights and biases from the previous two three-layer autoencoders. Also initialize the other hyperparameters with the best values found during optimization process.
17. Obtain D_M-dimensional representations (secondary feature vectors) on $X'_{training}$, $X'_{validation}$, and $X'_{testing}$ by transformation using the saved weights and biases (shown in Fig. 4c) (hereafter referred to as $X''_{training}$, $X''_{validation}$, and $X''_{testing}$).
18. Apply the same steps to the sensory data and obtain D_S-dimensional representation (secondary feature vector) of sensory data (hereafter referred to as $Y''_{training}$, $Y''_{validation}$, and $Y''_{testing}$) (shown in Fig. 7). Save the weights and biases of the neural network as W_3, W_4, W_3', W_4', $\mathbf{b}_3$, $\mathbf{b}_4$, $\mathbf{b}_3'$, and $\mathbf{b}_4'$, respectively.

3.2 Mapping Feature Vectors

Build and train a neural network for mapping between two types of secondary feature vectors, the input and the target of which are secondary feature vectors of mass spectra X'' and secondary feature vectors of sensory data Y''. Although these low-dimensional representations are incomprehensible to us in most cases, they can be decoded to data of original dimensions with the saved weight and biases.

1. Build a five-layer neural network. From the input layer to the output layer, each layer has D_M, K_{P1}, K_{P2}, K_{P3}, and D_S neurons, respectively (Fig. 8).
2. Choose one sample randomly from $X''_{training}$ and calculate the output of the neural network.
3. Calculate the error between the output of the neural network and the target signal in $Y''_{training}$.
4. Update the weights and the biases with the back-propagation algorithm.

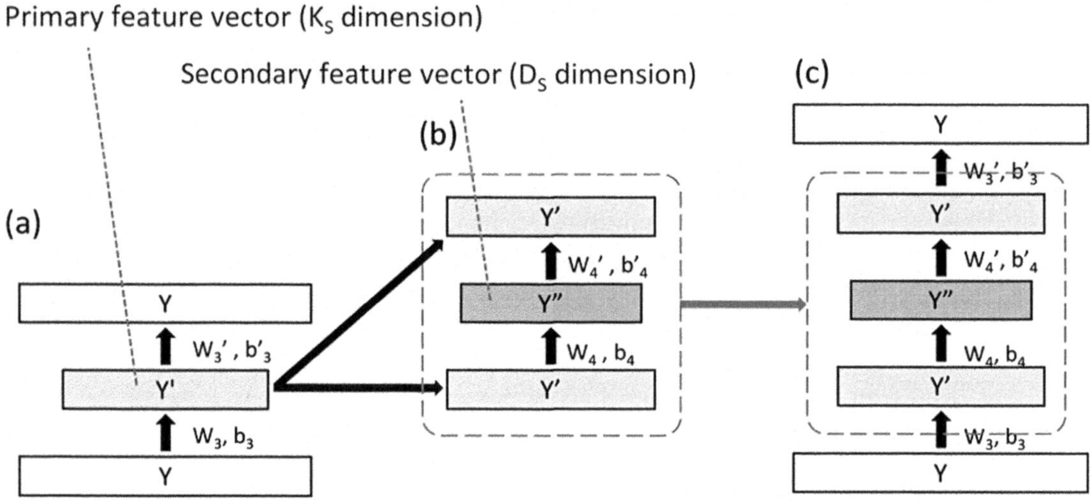

Dimensionality		
144-K_S-144	K_S-D_S-K_S	144-K_S-D_S-K_S-144

Fig. 7 A diagram of autoencoders (**a**) A diagram of a three-layer autoencoder compressing the original dataset into K_S dimension. (**b**) A diagram of a three-layer autoencoder compressing the K_S-dimensional dataset into D_S dimension. (**c**) A diagram of a five-layer autoencoder compressing the original dataset into D_S dimension

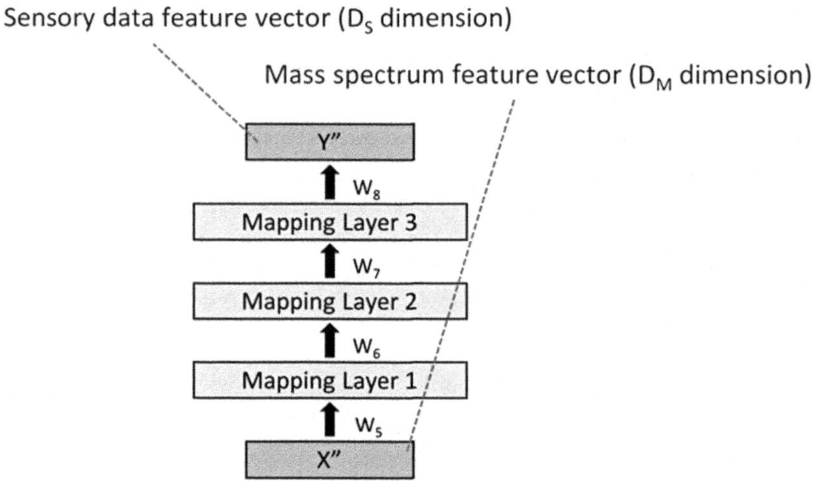

Dimensionality
D_M-K_{p1}-K_{p2}-K_{p3}-D_S

Fig. 8 A diagram of a five-layer neural network mapping mass spectra feature space to sensory data feature space

5. Choose one sample randomly from $X''_{validation}$ and calculate the output of the neural network with the updated weights and biases.
6. Calculate the error between the output of the neural network and the target signal $Y''_{validation}$.
7. Repeat **steps 2–6** until the error at **step 6** converges.
8. Optimize hyperparameters to obtain less error on $Y''_{validation}$. Optimization is done by repeating **steps 2–7** while changing one of the hyperparameters.
9. Save the weights and biases of the neural network as W_5, W_6, W_7, W_8, $\mathbf{b}_5$, $\mathbf{b}_6$, $\mathbf{b}_7$, and $\mathbf{b}_8$, after training. Also save the other hyperparameters.

3.3 Predictive Model

Combine the three neural networks to form a nine-layer predictive neural network, and perform a tuning on the model to improve its accuracy. The first three layers are taken from the autoencoder used on mass spectral data to transform vector representations of mass spectra into their secondary feature vectors. The middle three layers are taken from the mapping neural network to convert the secondary feature vectors of mass spectra into secondary feature vectors of sensory data. The last three layers are taken from the autoencoder used on sensory data to transform secondary feature vectors of sensory data to vector representations of 144-dimensional sensory data.

1. Build a nine-layer neural network. From the input layer to the output layer, each layer has 212, K_M, D_M, K_{P1}, K_{P2}, K_{P3}, D_S, K_S, and 144 neurons, respectively (Fig. 9).
2. Initialize hyperparameters with the ones which were found during procedures 3.1 and 3.2. Initialize the weights and the biases with the saved weights and biases W_1, W_2, W_5, W_6, W_7, W_8, W_4', W_3',$\mathbf{b}_1$, $\mathbf{b}_2$, $\mathbf{b}_5$, $\mathbf{b}_6$, $\mathbf{b}_7$, $\mathbf{b}_8$, $\mathbf{b}_4'$, and $\mathbf{b}_3'$. For training, validation, and testing data of target signal, $Y_{training}$, $Y_{validation}$, and $Y_{testing}$ are used, respectively.
3. Choose one sample randomly from $X_{training}$ and calculate the output of the neural network.
4. Calculate the error between the output of the neural network and the target signal in $Y_{training}$.
5. Update the weights and the biases with the back-propagation algorithm.
6. Choose one sample randomly from $X_{validation}$ and calculate the output of the neural network with the updated weights and biases.

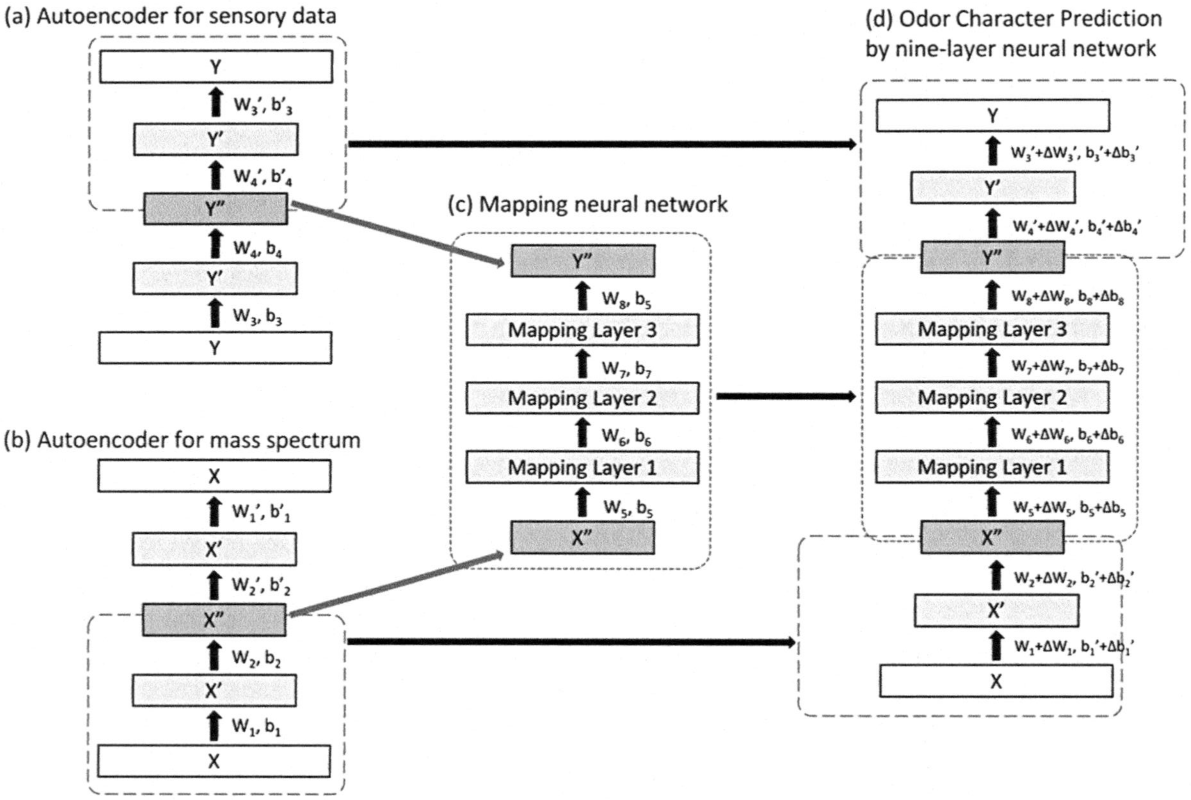

Dimensionality		
(a) 144-K_S-D_S-K_S-144 (b) 212-K_M-D_M-K_M-212	(c) D_M-K_{p1}-K_{p2}-K_{p3}-D_s	(d) 212-K_M-D_M-K_{P1}-K_{P2}-K_{P3}-D_S-K_S-144

Fig. 9 A diagram of predictive modeling. (**a**) A five-layer autoencoder for dimensionality reduction of sensory data. (**b**) A five-layer autoencoder for dimensionality reduction of mass spectra data. (**c**) A five-layer neural network mapping mass spectrum feature space to a sensory data feature space. (**d**) A nine-layer neural network for odor character prediction

7. Calculate the error between the output of the neural network and the target signal.
8. Repeat **steps 3–7** until the error on $Y_{validation}$ converges.
9. Optimize hyperparameters to obtain less error on $Y_{validation}$. Optimization is done by repeating **steps 3–8** while changing one of the hyperparameters.
10. Save the weights and biases of the neural network as $W_1 + \Delta W_1$, $W_2 + \Delta W_2$, $W_5 + \Delta W_5$, $W_6 + \Delta W_6$, $W_7 + \Delta W_7$, $W_8 + \Delta W_8$, $W_4' + \Delta W_4'$, $W_3' + \Delta W_3'$, $b_1 + \Delta b_1$, $b_2 + \Delta b_2$, $b_5 + \Delta b_5$, $b_6 + \Delta b_6$, $b_7 + \Delta b_7$, $b_8 + \Delta b_8$, $b_4' + \Delta b_4'$, *and* $b_3' + \Delta b_3'$, respectively. Also save the other hyperparameters.

3.4 Performance Evaluation

Evaluate the performance of the nine-layer predictive model on unknown samples. There are several possible methods to evaluate the performance of a predictive model, such as Euclidean norm, relative error, and correlation coefficient between a pair of a predicted vector and an actual vector. In our previous simulation, we used the correlation coefficient for the performance evaluation (*see* **Note 14**), and obtained correlation coefficient of *ca.* 0.76 between outputs and target signals on the testing set. Figure 10 shows a scatterplot of the odor character of one chemical predicted by a nine-layer neural network versus the target signal. Cross-validation is usually used for performance evaluations (*see* **Note 15**).

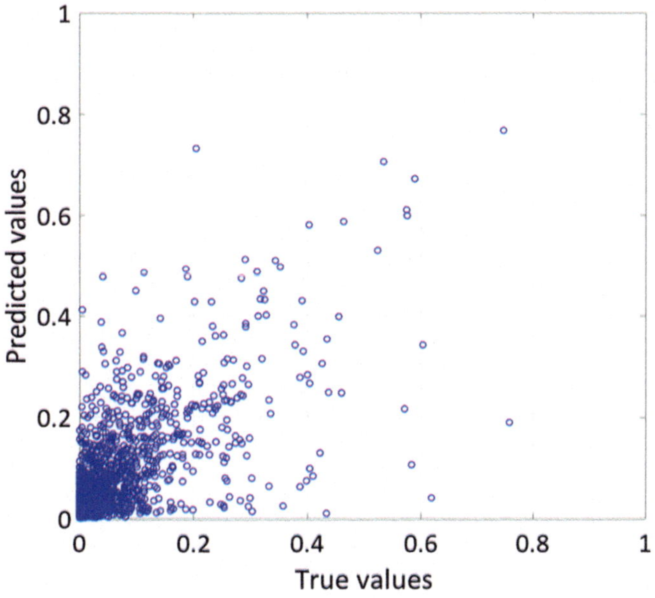

Fig. 10 Example of prediction

4 Notes

1. Neural networks are inspired by the biological neural networks of animal brains [8]. Herein, we use feedforward neural networks, composed of several layers made of a collection of artificial neurons. As the name implies, in feedforward networks, information flows through the input to output while having no feedback connections.

   One single neuron has a D-dimensional input vector and has one single output signal.

   The schematic of one neuron is shown in Fig. 11. The response of the neuron is typically a nonlinear function of the weighted sum of all inputs. In the following notes, the constant bias in a feedforward neural network is assumed to be included in the input vector $\boldsymbol{x}$. A weight vector W is used to weight the individual input signals. Then the response of the single neuron y is

$$y = f\left(\sum_{d=1}^{D+1} W_d x_d\right) \tag{1}$$

   The activation function $f(\cdot)$ is a nonlinear function such as the logistic sigmoid function (*see* **Note 9**).

   Multilayer feedforward neural networks consist of multiple layers of neurons. The first layer corresponds to input (and bias) signals, and other layers in the multilayer network receive their inputs only from all neurons in the previous layer and from one bias signal. The output signals of multilayer feedforward neural network are the output signals of the neurons in the last layer. Figure 12 shows a three-layer feedforward neural network. The input vector of the neural network is the input vector of the neurons in the hidden layer.

2. The results of the sensory evaluation carried out by Dravnieks [6] in which 160 odorants were evaluated for each of 146 verbal descriptors on a scale of 0–5 are used for the target dataset.

3. Mass specta of chemicals are used as input data. All mass spectra were obtained with electronic ionization method with energy

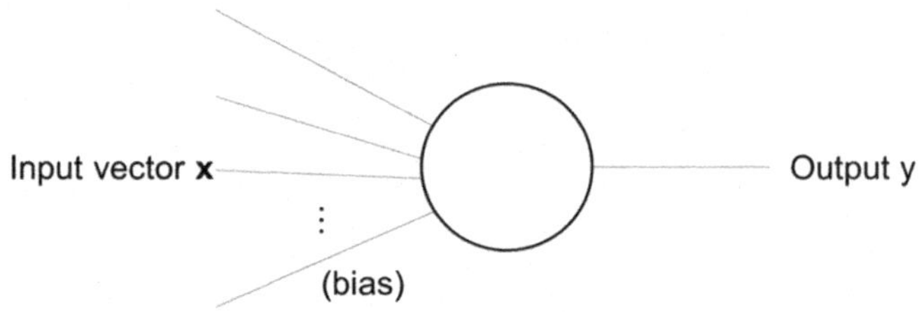

Fig. 11 A diagram of single neuron

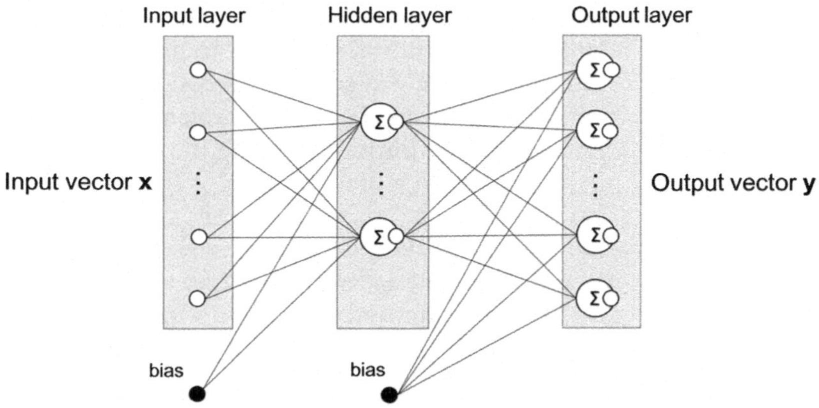

Fig. 12 A diagram of a three-layer feedforward neural network

of 70 [eV]. They were obtained from the Chemistry WebBook provided by National Institute of Standards and Technology. After taking the chemicals listed in both datasets, we obtained 121 samples for our experiments.

4. Normalization is fundamentally necessary to make the dataset to fit a neural network. There are several ways for normalization: (a) Divide all the elements in the matrix by the maximum value in the dataset, and (b) divide the elements in each row (sample vector) by the maximum value in the row. The former normalization keeps both the shape of the mass spectra and the magnitude relations among the samples while the latter normalization keeps only a shape of mass spectra. In this chapter, we applied the former for dataset normalization.
5. Intensities between 51 and 262 m/z are extracted from the original dataset since intensities at m/z below 50 mainly originate from odorless molecules such as oxygen, nitrogen, and carbon dioxide, and intensities at high m/z originate from molecules with low volatility and little effect on odor characteristics [9].
6. We omitted two descriptors since they have a correlation coefficient of 1.00 to other descriptors.
7. Autoencoders are a special family of artificial neural networks, the purpose of which is to learn from a compressed representation from a set of data [10]. An autoencoder consists of an input layer, one or more hidden layer(s), and an output layer with the same number of neurons as the input layer. Since fewer neurons are set in the middle hidden layer, a low-dimensional representation can be obtained. A projection function to a low-dimensional representation is acquired through the training process [11].

8. Dimensionality reduction, also called feature extraction, is a common technique often applied to neural networks to accomplish an effective projection function while avoiding the problems arising from the higher dimensionality. Increasing the dimensionality without increasing the number of training samples will result in a deterioration of the model's performance. This problem is known as "*the curse of dimensionality*" [12]. Dimensionality reduction also affords the reduction of noise from a dataset and computational cost. Therefore, this technique is commonly used in a variety of machine learning applications [13].

9. Activation functions can be a variety of shapes. Well-known activation functions are sigmoid functions, logistic functions, hyperbolic tangents, or rectifiers [14].

   A sigmoid function is defined as

$$f(x) = \frac{1}{1 + e^{-x}} \tag{2}$$

10. Back-propagation [15] is the most common learning algorithm. Errors are back-propagated from output to input. Optimization of the weights is achieved by minimizing an error function (also called cost function). Mean squared error is probably the most widely used error function. If the output vector of the neural network is $\hat{y}$ and the target vector is y, the error function is defined as

$$E = \frac{1}{2} \sum_{d} (y_d - \widehat{y_d})^2 \tag{3}$$

   where the error is summed over the neurons in the output layer. There are several different common error functions to be chosen. The cross-entropy loss is also a commonly used error function. We also previously reported the Itakura-Saito distance as an alternative error function for a neural network [16].

   The minimization of the error function is done using gradient-descent method. $w_{(t)kj}$ represents the weights of the output layer at the tth epoch, $w_{(t-1)kj}$ represents the weights of the output layer at the $(t-1)$th epoch, and z_j represents the activation of the jth neuron of the hidden layer, that is,

$$w_{(t)kj} = w_{(t-1)kj} + \Delta w_{kj}, \tag{4}$$

$$\Delta w_{kj} = \eta \frac{\partial E}{\partial w_{kj}}, \tag{5}$$

   and

$$\frac{\partial E}{\partial w_{kj}} = (y_k - \widehat{y_k}) z_j, \tag{6}$$

where η is a learning rate, a positive constant used to control the learning speed, k is the index of the neuron of the previous layer, and j is the index of the neuron in the output layer. The delta rule is the most common technique to calculate the derivative of E with respect to weight between input and hidden layers. For a three-layer network where x_i represents the activation of the ith neuron of the input, y_k represents the activation of the kth neuron of the output layer, w_{kj} is the weights of the output layer between y_k and z_j, and v_{ji} is the weights of the hidden layer between z_j and x_i; the rule is

$$\frac{\partial E}{\partial v_{ji}} = \sum_k \sum_j \frac{\partial E}{\partial y_k} \frac{\partial y_k}{\partial z_j} \frac{\partial z_j}{\partial \sum_i v_{ji} x_i} \frac{\partial \sum_i v_{ji} x_i}{\partial v_{ji}} \tag{7}$$

Rewriting this equation with Eqs. 2 and 3 gives

$$= \sum_k \sum_j (y_k - \hat{y_k}) w_{kj} z_j (1 - z_j) x_i. \tag{8}$$

If the neural network has more than one hidden layer, weights of the previous hidden layer can be computed with Eq. (8) in the same manner.

Stochastic gradient descent:
Instead of using full training set to compute Δw (this is called the batch method), the stochastic gradient method [17] uses one sample which is randomly taken from the training set to compute next Δw. While the batch method tends to be very slow and requires a large machine memory, stochastic gradient descent can overcome these problems and provide faster convergence.

Momentum:
The momentum method [18] is a technique for accelerating gradient descent that accumulates a velocity vector in directions of persistent reduction. In this technique, velocity v is added to Eq. 4:

$$v_{(t)} = \alpha v_{(t-1)} - \eta \frac{\partial E}{\partial w_{kj}} \tag{9}$$

where α is a positive constant usually set to 0.9 or a similar value, $v_{(t)}$ represents the next velocity, and $v_{(t-1)}$ represents the current velocity, initialized with 0.

Regularization:
Regularization [19] is used to prevent overfitting. It uses an additional term in an error function which is typically chosen to impose a penalty on the complexity of the model. The usual expression of L2 norm penalty term, based on the mean

squared error, is extended by adding a regularization term to the error. When calculating error at the output layer, Eq. 3 is rewritten as

$$E = \frac{1}{2}\sum_{k}\left(y_k - \widehat{y_k}\right)^2 + \frac{\lambda}{2}\sum_{k}\sum_{j}\|w_{kj}\|^2, \tag{10}$$

where λ is a positive constant to control the regularization term. The dropout method is also a well-known method for regularization [20].

11. To train a deep autoencoder, a layer-wise approach is one of the most common techniques [20]. In this approach, each layer is trained in turn. First, train the first layer on raw data to obtain weights and biases which transform the raw data into a primary feature vector. Then train the second layer on this feature vector to obtain weights and biases which transform primary feature vectors into secondary feature vectors.
12. Normally 70% of the available data is allocated for training. The remaining 30% of the data is equally partitioned and referred to as validation and test datasets. The training set is used to learn the weights and the biases. The validation set is used to avoid overfitting. If the error on this validation set is increasing during the training process, it indicates that the neural network is overfitting to training data and then you should stop training. The testing set is used to assess the general ability of the constructed model.
13. Weight initialization can have an impact on both the convergence rate and the performance of the model. Although there are many different initialization methods based on different ideas, Xavier initialization is widely used in many deep learning frameworks. In Xavier weight initialization [21], where a neuron has d connections from the neurons in the previous layer, initialize weights by Gaussian distribution with the standard deviation of $(1/\sqrt{d})$.
14. To compute the correlation coefficient R between predicted values and true values, where $\widehat{y_n}$ represents the predicted vector of the nth sample in the testing set, $y_n$represents the true vector of nth sample in the testing set, $Cov(\widehat{y_n}, y_n)$ is a covariance between $\widehat{y_n}$ and y_n, and $\sigma_{\hat{y}}$ and σ_y are the standard deviations of $\widehat{y_n}$ and y_n, respectively:

$$R = \sum_{n}\frac{Cov(\widehat{y_n}, y_n)}{\sigma_{\hat{y}}\sigma_y} \tag{11}$$

15. In cross-validation, run this modeling process on different subsets of the data to obtain multiple measures of model performance. For example, in fivefold cross-validation, divide the data into five subsets of 20% of the full dataset.

Acknowledgments

This work was partly supported by JST (Japan Science and Technology Agency)-Mirai Program, Grant Number JPMJMI17DD, Japan.

References

1. LeCun Y, Bengio Y, Hinton G (2015) Deep learning. Nature 521:436–444
2. Alan M (1995) Applications of neural networks. Kluwer Academic Publishers
3. MacQueen J, (1967) Some methods for classification and analysis of multivariate observations, The Regents of the University of California.
4. Buck L, Axel R (1991) A novel multigene family may encode odorant receptors: a molecular basis for odor recognition. Cell 65:175–187
5. Nozaki Y, Nakamoto T (2016) Odor impression prediction from mass spectra. PLoS One 11:e0157030
6. Dravnieks (1992) A. Atlas of odor character profiles
7. NIST Chemistry WebBook (2017). http://webbook.nist.gov/chemistry/. Accessed 10 August 2017
8. Jamse A (1995) An introduction to neural networks. The MIT Press, Cambridge
9. Nakamoto T, Ohno M, Nihei Y (2012) Odor approximation using mass spectrometry. IEEE Sens J 12:3225–3231
10. Hinton G (2007) Learning multiple layers of representation. Trends Cogn Sci 11:428–434
11. Hinton G, Salakhutdinov R (2006) Reducing the dimensionality of data with neural networks. Science 313:504–507
12. Donoho D. (2000) High-dimensional data analysis: the curses and blessings of dimensionality, The American Mathematical Society, pp. 1–33
13. Lee J, Verleysen M (2007) Nonlinear dimensionality reduction. Springer Science & Business Media
14. Nair V, Hinton G (2010) Rectified linear units improve restricted Boltzmann machines. In: Proceedings of the 27th International Conference on International Conference on Machine Learning, pp 807–814
15. Rumelhart D, Hinton G, Williams R (1986) Learning representations by back-propagating errors. Nature 323:533–536
16. Nozaki Y, Nakamoto T (2017) Itakura-Saito distance based autoencoder for dimensionality reduction of mass spectra. Chemom Intel Lab Syst 167:63–68
17. Wiegerinck W, Komoda A, Heskes T (1994) Stochastic dynamics of learning with momentum in neural networks. J Phys Math Gen 27:4425
18. Qian N (1999) On the momentum term in gradient descent learning algorithms. Neural Netw Off J Int Neural Netw Soc 12:145–151
19. Tibshirani R (1996) Regression shrinkage and selection via the Lasso. J R Stat Soc Ser B Methodol 58:267–288
20. Srivastava N, Hinton G, Krizhevsky A, Sutskever I, Salakhutdinov R (2014) Dropout: a simple way to prevent neural networks from overfitting. J Mach Learn Res 15:1929–1958
21. Glorot X, Bengio Y (2010) Understanding the difficulty of training deep feedforward neural networks, PMLR 249–256.

Chapter 4

A Photochromic Sensor Microchip for High-Performance Multiplex Metal Ion Detection

Meng Qin, Fengyu Li, and Yanlin Song

Abstract

Photochromic molecules can respond to external stimulations and undergo reversible conversion between different chemical structures, providing one photochromic molecule with multiple recognition states for targeting compounds. Here we design a facile sensor microchip with only one photochromic molecule (spirooxazine) to discriminate multiplex metal ions. The sensor chip performs in dark, ultraviolet, or visual stimulation, resulting in different molecular states of spirooxazine-metallic coordination and patterned fluorescent signals for analysis. By using this sensor microchip, 11 metal ions are discriminated. Furthermore, mineral water of 16 different brands and metal ions in human serum are distinguished.

Key words Photochromic sensor, Spirooxazine, Metal ions, Multistates, Fluorescence, Multivariate analysis

1 Introduction

Multi-analyte recognition has drawn increased attention in the areas of environmental monitoring, clinical diagnosis, biological screening, food industry, etc. [1–11]. The mammalian olfactory system has inspired the development of cross-reactive sensor arrays for targeting multi-analyte recognition. The key feature of a cross-reactive sensor array is that the nonspecific sensing elements can respond differentially to all the analytes, thus providing abundant sensing information and distinguished fingerprint signal for each analyte [12, 13]. Generally, to construct a cross-reactive sensor array, large numbers of sensing molecules or indicators have to be employed to attain sufficient sensing information; however, this usually involves complicated chemical synthesis or massive screening. For example, Anzenbacher et al. performed multi-analysis of carboxylate drugs by synthesizing a series of sensors [14]; Anslyn et al. constructed indicator displacement sensor arrays to discriminate ginsenosides and ginsengs with screening a suite of indicators [15]. To achieve high-efficiency and low-cost recognition of

Jessica E. Fitzgerald and Hicham Fenniri (eds.), *Biomimetic Sensing: Methods and Protocols*, Methods in Molecular Biology, vol. 2027, https://doi.org/10.1007/978-1-4939-9616-2_4,

multiple analytes, it is highly desired to design sensor arrays with a minimum number of sensing elements.

Spiropyran and spirooxazine (SP) are a class of photochromic molecules, which can undergo interconversion of chemical structures between ring-closed and ring-opened merocyanine (MC) form by responding to external stimuli such as light, proton, and metal ions [16–21]. The nitrogen and oxygen atoms in the oxazine ring of SP are excellent binding sites through supermolecular interaction, while the charged MC state can combine with various molecules [22–24]. Thus, an SP molecule alone can act as a multi-state receptor for diverse analytes, demonstrating its potential as a cross-reactive sensing element with abundant chemical information.

Here, we adopt a commercial SP, 1,3,3-trimethylindolinonaphthospirooxazine (TNSP), to design a sensor microchip for targeting multiple metal ions through fluorescence detection (Fig. 1). With manipulation of external light stimulations (dark, ultraviolet (UV) irradiation, and visible (Vis) irradiation), TNSP can undergo a reversible conversion of chemical structures between the ring-closed state and ring-opened MC state. The merocyanine spirooxazine can combine with proton or metal cations to form metallic merocyanine (MMC), as shown in Fig. 2. Performing the reversible reactions between closed, MC, and MMC forms of TNSP, different ion strengths and light stimulations have a differential inductive effect, which leads to fluorescent changes of TNSP (Fig. 3). This photochromic sensor microchip can actualize detections of 11 various metal ions with only one facile sensor, which offers a new strategy to design and develop a

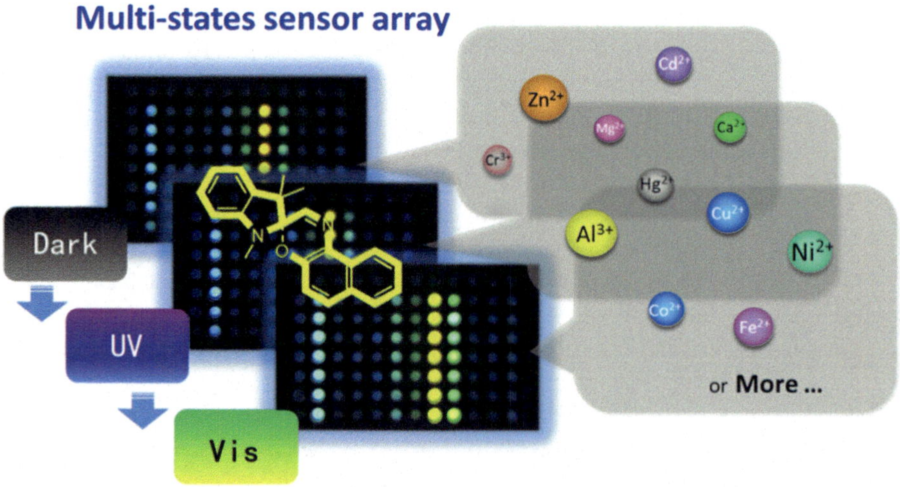

Fig. 1 The schematic illustration of the photochromic multistates sensing for multiplex metal ion detection with one spirooxazine sensor. Three states (dark, UV irradiation, and Vis irradiation) of spirooxazine-metallic coordination can be obtained by spirooxazine response with metal ions in different irradiation conditions (reproduced from ref. 22 with permission from Macmillan Publishers Limited)

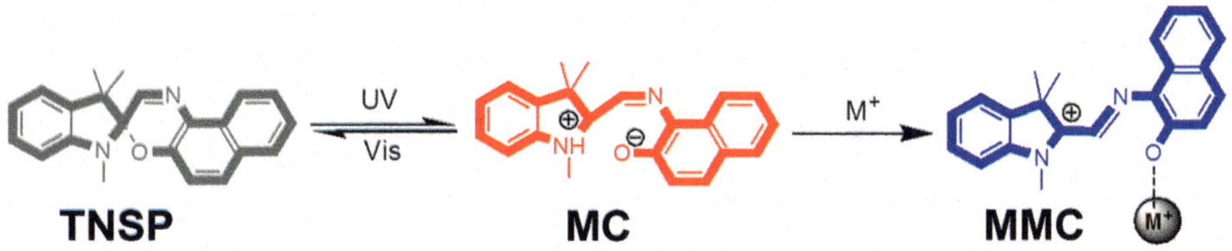

Fig. 2 The interconversion of TNSP among the closed form, merocyanine (MC) form, and metallic merocyanine (MMC) form (reproduced from ref. 22 with permission from Macmillan Publishers Limited)

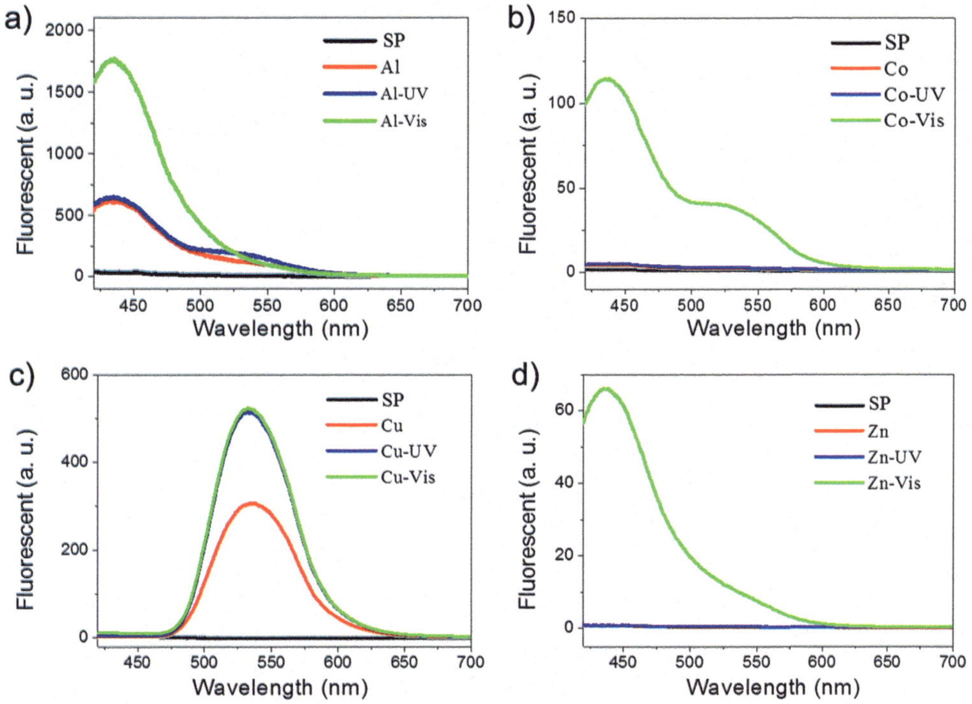

Fig. 3 The fluorescence of TNSP responding to various metal ions with different light irradiations. The fluorescence of TNSP at 435 nm or 533 nm is changed with response to (**a**) Al^{3+}, (**b**) Co^{2+}, (**c**) Cu^{2+}, and (**d**) Zn^{2+} in dark, UV, and Vis irradiations (reproduced from ref. 22 with permission from Macmillan Publishers Limited)

high-performance cross-reactive sensor array. Furthermore, for practical applications, mineral water of 16 different brands and metal ions in human serum can be distinguished.

2 Materials

2.1 Photochromic Sensor Microchip

1. Micro-well plates (1000 μm wide, 250 μm deep).
2. Ethanol solution of thermoplastic polyurethane Tecoflex® with concentration of 0.5 wt% (*see* **Note 1**).
3. TNSP is dissolved in the above solution (in dark condition) with a concentration of 1.0 mM (*see* **Note 2**).

2.2 Solutions of Target of Interest

1. Aqueous solutions of 11 metal chlorides: $AlCl_3$, $CaCl_2$, $CdCl_2$, $CoCl_2$, $CrCl_3$, $CuCl_2$, $FeCl_2$, $HgCl_2$, $MgCl_2$, $NiCl_2$, and $ZnCl_2$, with a concentration of 1.0 mM and pH = 5 (*see* **Note 3**).
2. Control sample of HCl–NaCl aqueous solution: Concentration of NaCl is 1.0 mM, pH = 5.
3. Sixteen different brands of mineral and purified water: Baby Bella, FIJI, Nongfu Spring, *S. Antonio*, San Benedetto, Sourcy Pure Red, Teleno, Vswp, Evian, Highland, Perrier, Sam Da Soo, Sourcy Naturelle, SPA, Volvic, and Waiwera (*see* **Note 4**).
4. Metal ions dissolved in human serum: Ethanol is added into human serum (2.5/1 v/v), centrifuged at high speed (11,180 × *g*, 5 min), then metal chlorides added into the supernatant to prepare 1.0 mM solutions (*see* **Note 5**).

2.3 Equipment of Illumination Source and Data Collection

1. UV LED lamp (wavelength of 245 nm) as the UV illumination source.
2. Solar simulator as the Vis illumination source.
3. Fluorescence scanner (ChampChemi Professional+) in six channels: CH1: 450 nm, CH2: 480 nm, CH3: 505 nm, CH4: 535 nm, CH5: 570 nm, and CH6: 605 nm, with 365 nm UV light excitation (*see* **Note 6**).

3 Methods

3.1 Fabrication of TNSP Microchip with Metal Ions

1. Pipette TNSP/Tecoflex® solution into micro-wells of the plate to form a 12 × 7 matrix, with 200 nL solution for each well. After drying the solvent, the TNSP microchip is formed.
2. Pipette metal ion solutions and control solution onto the microchip, with 200 nL solution for each pixel (micro-well) and seven trials for each sample.

3.2 Conversion of Sensor States by Manipulating Light Stimulations

1. Initial reaction is performed in the dark within 10 min.
2. After collecting the fluorescence responses, place the sensor microchip under UV irradiation for 10 min to trigger the structural conversion of TNSP to MC and MMC forms.
3. After collecting the fluorescence responses, place the sensor microchip under Vis radiation for 10 min to trigger the reversible structural conversion of TNSP.

3.3 Data Collection and Analysis

1. Collect the fluorescence responses of the microchip to different metal ions under dark, UV irradiation, and Vis irradiation, respectively, by fluorescence scanner.

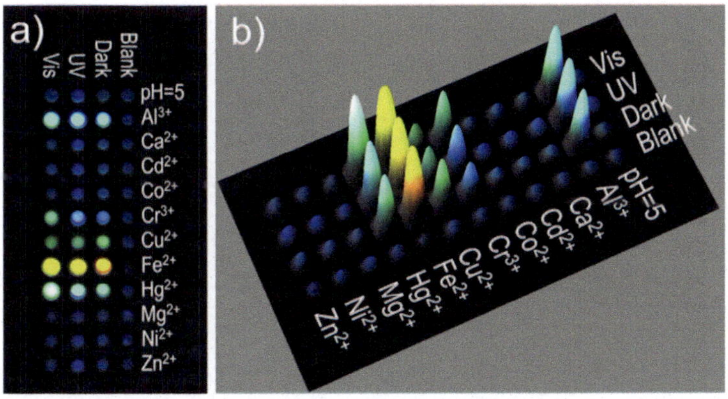

Fig. 4 (**a**) Fluorescent image of the photochromic TNSP microchip responding to metal ions under different irradiation conditions. (**b**) The 3D representation of the integrated fluorescence intensity corresponding to the fluorescent image (reproduced from ref. 22 with permission from Macmillan Publishers Limited)

2. Superimpose the equally weighted images corresponding to red, green, and blue (RGB) channels. The fluorescent image of the microchip is shown in Fig. 4. The data processing consists of the integration of the fluorescence intensity per pixel of the TNSP microchip before (blank) and after the solutions of metal ions are spotted (*see* **Note** 7).
3. Utilize linear discriminant analysis (LDA) and hierarchical clustering analysis (HCA) to carry out the multivariate analysis. The results are shown in Fig. 5 (*see* **Notes 8** and **9**).
4. For the sensing of potable water samples listed in Table 1, the procedure is same as that for metal ions. No pretreatment is carried out on the water samples.
5. For the sensing of metal ions in human serum, the procedure is same as that for aqueous solutions of metal ions.

3.4 Sensing Application

The TNSP-metallic complexes can be further utilized to construct a photochromic sensor chip for the recognition of 20 natural amino acids, i.e., Ala, Arg, Cys, Gly, Lys, Pro, Met, Thr, Ser, His, Val, Tyr, Asn, Glu, Trp, Gln, Ile, Phe, Asp, and Leu. Upon addition of amino acids, the coordination between TNSP-metallic complex and amino acids facilitates the formation of a new balance; different amino acids lead to diverse balances with various fluorescence. The working principle is shown in Fig. 6.

1. Recognition of 20 natural amino acids in phosphate-buffered saline (PBS) solution. The photochromic sensor chip is composed of ten TNSP-metallic sensing complexes (Al^{3+}-, Ca^{2+}-, Cd^{2+}-, Co^{2+}-, Cu^{2+}-, Fe^{2+}-, Hg^{2+}-, Mg^{2+}-, Ni^{2+}, and Zn^{2+}-based TNSP-metallic complexes). Fluorescent sensing information is collected from dark, UV, and Vis conditions. The

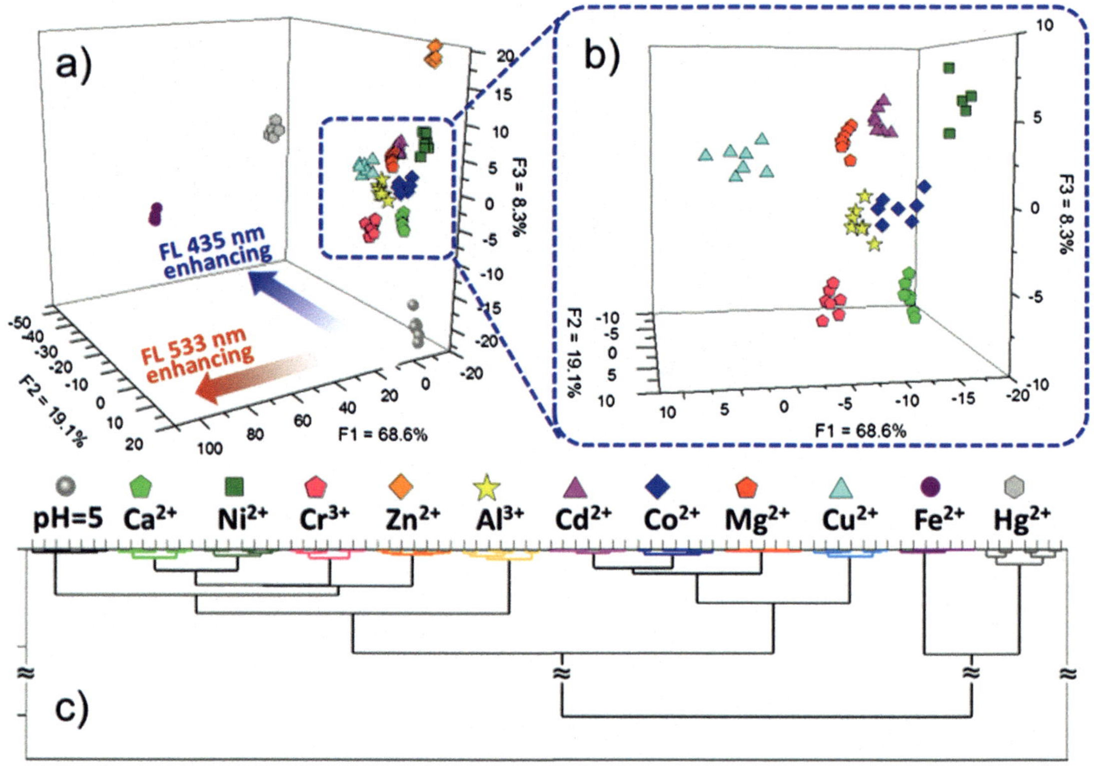

Fig. 5 (**a**, **b**) LDA score plot of the detection result and its corresponding magnified image. 11 clusters of metal ions are clearly separated. LDA reflects analyte-specific fluorescent enhancement at 435 nm or 533 nm due to various metal electropositivity. (**c**) HCA gives the similarity clustering of the analytes based on the fluorescent variation trend of the TNSP-metallic coordination in three states (reproduced from ref. 22 with permission from Macmillan Publishers Limited)

Table 1
Metal ion content for different brands of mineral water samples

Metal ion content (mg/L)							
Mineral water	**Ca**	**Mg**	**pH**	**Mineral water**	**Ca**	**Mg**	**pH**
Baby Bella	53.3	14.9	–	Evian	80	26	7.2
FIJI	17.5	–	7.7	Highland	40	10.1	7.8
Nongfu Spring	40	5	7.3	Perrier	160	4.2	–
S. Antonio	32.4	5.3	7.9	Sam Da Soo	2.9	1.9	7.7
San Benedetto	50.3	30.8	7.52	Sourcy Naturelle	54	4.5	7.5
Sourcy Pure Red	54	4.5	4.7	SPA	4.5	1.3	6
Teleno	6	1.2	–	Volvic	11.5	8	7
Vswp	45	51	–	Waiwera	12	2.6	7.6

The samples contain different kinds and concentrations of metal ions (reproduced from ref. 22 with permission from Macmillan Publishers Limited)

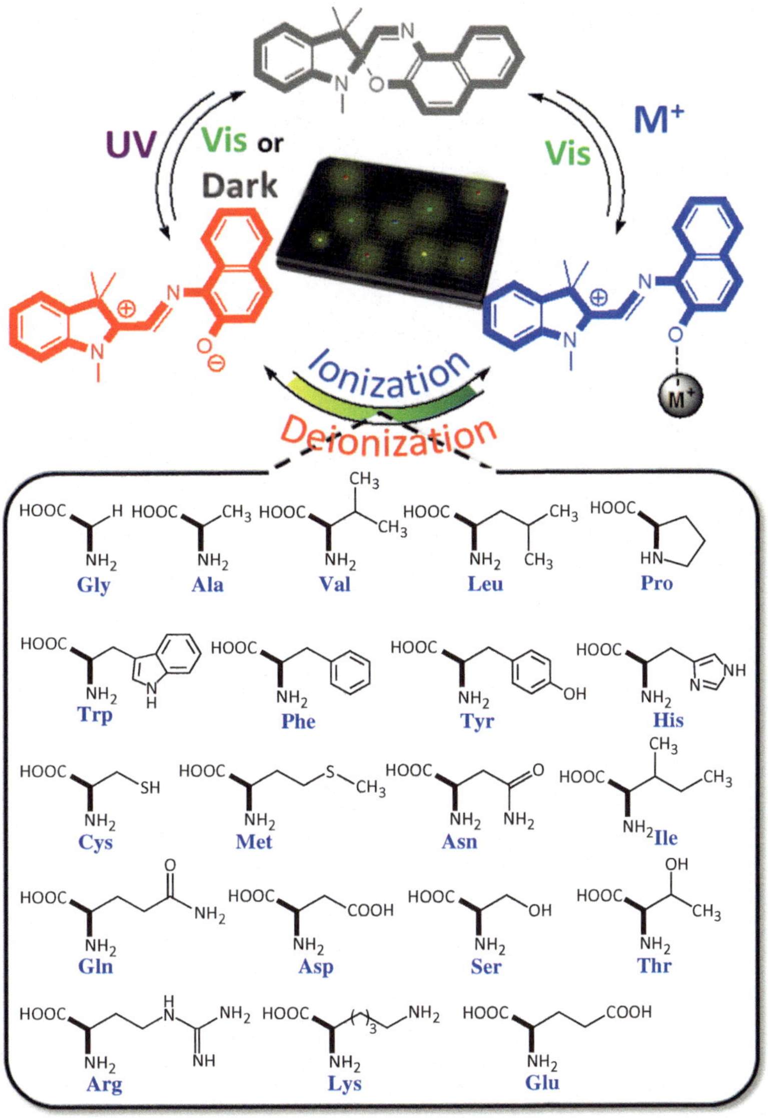

Fig. 6 Scheme of 20 natural amino acid identification by photochromic sensor chip composed of TNSP-metallic complexes. Under dark, UV, and Vis conditions, the sensor chip interconverts among different states. TNSP-metallic complex combines with amino acids and generates new balances with various fluorescence (reproduced from ref. 24 with permission from American Chemical Society)

result of the discriminant analysis is shown in Fig. 7. The concentrations of TNSP, metal ions, and amino acids are 0.5 mM, 0.5 mM, and 1.0 mM, respectively.

2. Recognition of 20 natural amino acids in human serum: The identification of amino acids can be successfully processed in human serum. The photochromic sensor chip is composed of

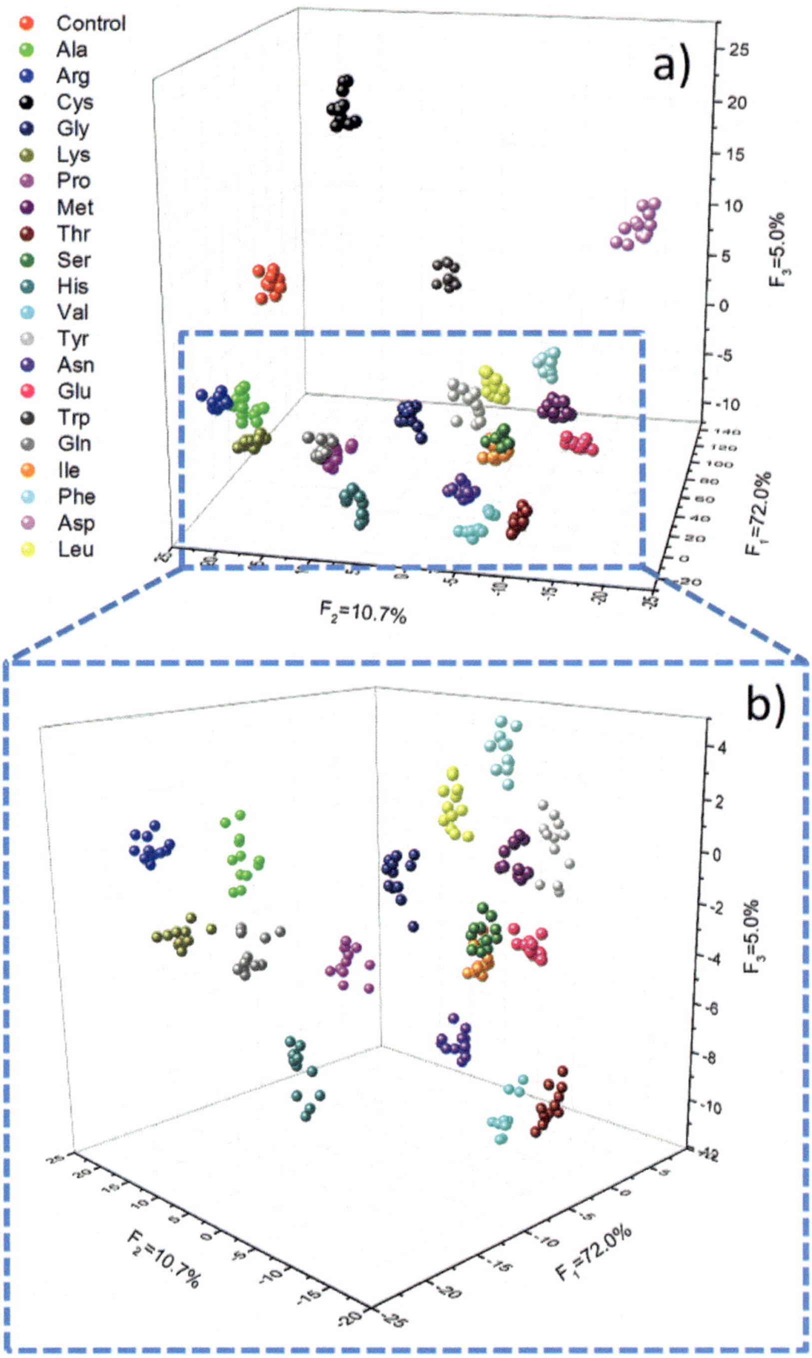

Fig. 7 Fluorescent discriminant analysis of 20 natural amino acids on photochromic TNSP-metallic sensor chip and rational analysis. (**a**) Graph of LDA result shows a clear clustering of the 20 natural amino acids and 1 PBS sample as control. (**b**) Corresponding magnified image. The sensor chip is composed of TNSP-metallic complexes involving metal ions as Al^{3+}, Ca^{2+}, Cd^{2+}, Co^{2+}, Cu^{2+}, Fe^{2+}, Hg^{2+}, Mg^{2+}, Ni^{2+}, and Zn^{2+} (reproduced from ref. 24 with permission from American Chemical Society)

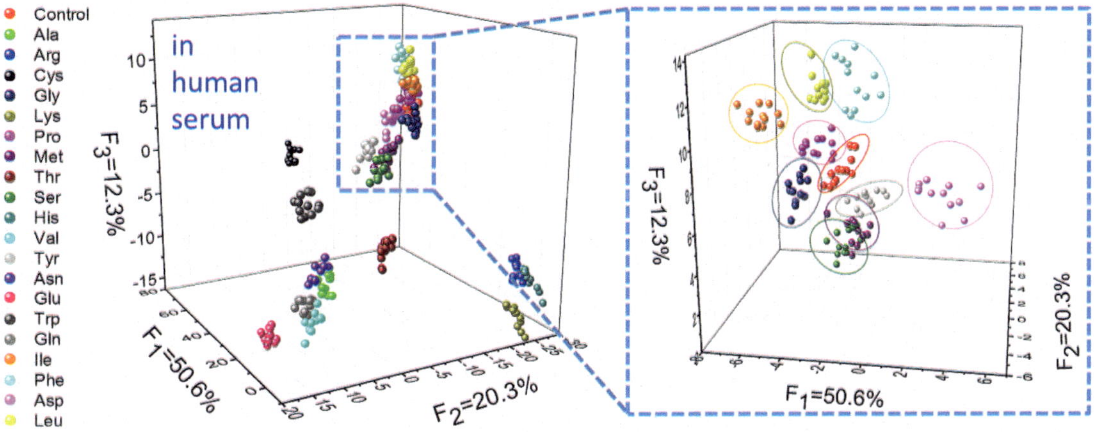

Fig. 8 Twenty natural amino acid discriminant analysis in human serum by the photochromic sensor chip consisting of just three metal ion (Fe^{2+}, Hg^{2+}, and Cu^{2+})-based TNSP-metallic complexes. Graph of LDA result and magnified image show separated clustering of the 20 natural amino acids in human serum (reproduced from ref. 24 with permission from American Chemical Society)

three metal ion (Fe^{2+}, Hg^{2+}, and Cu^{2+})-based TNSP-metallic complexes. The result of discriminant analysis is shown in Fig. 8. The concentrations of TNSP, metal ions, and amino acids are 0.5 mM, 0.5 mM, and 1.0 mM, respectively.

4 Notes

1. Tecoflex® polymer carrier matrix can immobilize TNSP molecule to form a solid sensor array. In addition, the hydrophilic polyurethane can draw the liquid analyte into the sensor material and ensure the formation of homogeneous sensor films.
2. Light and heat irradiation may induce conversion of TNSP chemical structures. Hence, dissolve the ethanol solution of TNSP in ultrasound bath with foil paper covering the container to block light irradiation. Add ice or keep recycling water in the ultrasound bath to eliminate the influence of heat.
3. The solutions of metal ions are controlled to be relatively acidic to prevent metallic sedimentation.
4. The portable water samples are actually an aqueous mixture of various metal ions. The Ca^{2+} and Mg^{2+} contents and pH values of all the mineral water brands are listed in Table 1.
5. Human serum provides a complex sensing environment. After removing the proteins, the supernatant still contains diverse metal ions and small biomolecules.
6. The six channels are determined by corresponding optical filters, which cover the whole visible wavelength.

7. Two control samples, "blank" and "pH = 5," are set for comparison. "Blank" represents TNSP solution before adding metal ions, while "pH = 5" represents TNSP solution spotted with water.
8. LDA method, carried out by SYSTAT® v12.02.00, is used to evaluate the similarities between the data corresponding to the same cluster by introducing the group classification. It provides a graphic representation useful to gain an insight into the clustering of the response data, and to calculate classification accuracy. Since LDA trains the data to describe the best-fit parameters to separate different clusters, the distance of the cluster in spatial distribution reveals differential fluorescent signals of the metal ions.
9. HCA, carried out by Minitab® v16.1.1.0, performs dimensionality reduction analysis to investigate the similarity between analytes by separating them into clusters. The cluster is defined by Ward's (minimum variance) method, which takes into consideration the minimum amount of variance between the samples. The HCA result shows a dendrogram of Euclidean distance between 84 samples (12 samples $\times$ 7 trials) with Ward linkage. HCA graphical output displays three major groups, which indicates the fluorescent variation trend of the different states of TNSP-metallic coordination.

Acknowledgments

This work was supported by the National Nature Science Foundation (Grant Nos. 51203166, 51473172, and 51473173), 973 Program (Nos. 2013CB933004), and the "Strategic Priority Research Program" of the Chinese Academy of Sciences (Grant No. XDA09020000). The Chinese Academy of Science is gratefully acknowledged.

References

1. Askim JR, Li Z, LaGasse MK, Rankin JM, Suslick KS (2016) An optoelectronic nose for identification of explosives. Chem Sci 7 (1):199–206
2. Lin H, Jang M, Suslick KS (2011) Preoxidation for colorimetric sensor array detection of VOCs. J Am Chem Soc 133 (42):16786–16789
3. Carey JR, Suslick KS, Hulkower KI, Imlay JA, Imlay KRC, Ingison CK, Ponder JB, Sen A, Wittrig AE (2011) Rapid identification of bacteria with a disposable colorimetric sensing array. J Am Chem Soc 133(19):7571–7576
4. Miranda OR, Li X, Garcia-Gonzalez L, Zhu Z-J, Yan B, Bunz UHF, Rotello VM (2011) Colorimetric bacteria sensing using a supramolecular enzyme-nanoparticle biosensor. J Am Chem Soc 133(25):9650–9653
5. Bajaj A, Miranda OR, Kim I-B, Phillips RL, Jerry DJ, Bunz UHF, Rotello VM (2009) Detection and differentiation of normal, cancerous, and metastatic cells using nanoparticle-polymer sensor arrays. Proc Natl Acad Sci U S A 106(27):10912–10916
6. De M, Rana S, Akpinar H, Miranda OR, Arvizo RR, Bunz UHF, Rotello VM (2009)

Sensing of proteins in human serum using conjugates of nanoparticles and green fluorescent protein. Nat Chem 1(6):461–465

7. Rana S, Le NDB, Mout R, Saha K, Tonga GY, Bain RES, Miranda OR, Rotello CM, Rotello VM (2015) A multichannel nanosensor for instantaneous readout of cancer drug mechanisms. Nat Nanotechnol 10(1):65–69
8. Umali AP, LeBoeuf SE, Newberry RW, Kim S, Tran L, Rome WA, Tian T, Taing D, Hong J, Kwan M, Heymann H, Anslyn EV (2011) Discrimination of flavonoids and red wine varietals by arrays of differential peptidic sensors. Chem Sci 2(3):439–445
9. Diehl KL, Anslyn EV (2013) Array sensing using optical methods for detection of chemical and biological hazards. Chem Soc Rev 42 (22):8596–8611
10. Minami T, Esipenko NA, Akdeniz A, Zhang B, Isaacs L, Anzenbacher P (2013) Multianalyte sensing of addictive over-the-counter (otc) drugs. J Am Chem Soc 135(40):15238–15243
11. Minami T, Esipenko NA, Zhang B, Kozelkova ME, Isaacs L, Nishiyabu R, Kubo Y, Anzenbacher P (2012) Supramolecular sensor for cancer-associated nitrosamines. J Am Chem Soc 134(49):20021–20024
12. Huang Y, Li FY, Qin M, Jiang L, Song YL (2013) A multi-stopband photonic-crystal microchip for high-performance metal-ion recognition based on fluorescent detection. Angew Chem Int Ed 52(28):7296–7299
13. Qin M, Huang Y, Li YN, Su M, Chen BD, Sun H, Yong PY, Ye CQ, Li FY, Song YL (2016) A rainbow structural-color chip for multisaccharide recognition. Angew Chem Int Ed 55(24):6911–6914
14. Liu Y, Minami T, Nishiyabu R, Wang Z, Anzenbacher P (2013) Sensing of carboxylate drugs in urine by a supramolecular sensor array. J Am Chem Soc 135(20):7705–7712
15. Zhang X, You L, Anslyn EV, Qian X (2012) Discrimination and classification of ginsenosides and ginsengs using bis-boronic acid receptors in dynamic multicomponent indicator displacement sensor arrays. Chem—Eur J 18(4):1102–1110
16. Kopelman RA, Snyder SM, Frank NL (2003) Tunable photochromism of spirooxazines via metal coordination. J Am Chem Soc 125 (45):13684–13685
17. Raymo FM, Giordani S (2001) Signal processing at the molecular level. J Am Chem Soc 123 (19):4651–4652
18. Shao N, Jin J, Wang H, Zheng J, Yang R, Chan W, Abliz Z (2009) Design of bis-spiropyran ligands as dipolar molecule receptors and application to in vivo glutathione fluorescent probes. J Am Chem Soc 132 (2):725–736
19. Meng X, Zhu W, Guo Z, Wang J, Tian H (2006) Highly stable and fluorescent switching spirooxazines. Tetrahedron 62 (42):9840–9845
20. Zhang J, Zou Q, Tian H (2013) Photochromic materials: more than meets the eye. Adv Mater 25(3):378–399
21. Zhang J, Wang J, Tian H (2014) Taking orders from light: progress in photochromic bio-materials. Mater Horiz 1(2):169–184
22. Huang Y, Li FY, Ye CQ, Qin M, Ran W, Song YL (2015) A photochromic sensor microchip for high-performance multiplex metal ions detection. Sci Rep 5:9724
23. Qin M, Huang Y, Li FY, Song YL (2015) Photochromic sensors: a versatile approach for recognition and discrimination. J Mater Chem C 3(36):9265–9275
24. Qin M, Li FY, Huang Y, Ran W, Han D, Song YL (2015) Twenty natural amino acids identification by a photochromic sensor chip. Anal Chem 87(2):837–842

Chapter 5

Contact Printing of a Quantum Dot and Polymer Cross-Reactive Array Sensor

Vincent P. Schnee and Collin J. Bright

Abstract

The incorporation of organic polymeric materials into chemical sensors and electro-optic devices has the potential to greatly advance these fields. A major challenge to their incorporation is the fabrication of thin films due to their intolerance of thermal deposition methods and solvent compatibility challenges. Here, a method for contact printing of quantum dot (QD) and organic polymer (OP) composites for the production of thin-film chemical sensors is described. The method described here allows for the repeatable, low-cost, and relatively simple production of thin films of QD/OP composites for use in chemical sensor arrays by a dry transfer process of the polymer on an elastomer stamp to the sensor substrate.

Key words Array sensor, Quantum dot, Polymers, Contact printing, Solvent free

1 Introduction

Chemical sensing arrays designed to mimic the mammalian sense of smell have shown great promise for the detection and identification of a wide variety of chemicals. These devices rely on multiple sensors working in concert to create a unique response pattern for each impinging analyte [1, 2]. Quantum dots (QD) have been an effective sensing material in these types of chemical sensing arrays [3–5] in addition to their other uses as light emitters [6, 7], photodetectors [8, 9], and fluorescent labels [10, 11]. In this work, we demonstrate a cross-reactive sensing array comprised of QD/polymer composites that use an unmodified 480 nm CdSe QD incorporated within varying polymers to add chemical diversity.

Fabrication of the thin films, which are typically found in chemical sensing arrays, can be labor intensive and technically challenging. Sensor "pixels" are comprised of many different materials, which are often required to be spatially close to one another. With this many different materials, adjacent fabrication of such an array can be quite difficult. Quantum dots are incompatible with many common fabrication techniques and additional challenges are

Jessica E. Fitzgerald and Hicham Fenniri (eds.), *Biomimetic Sensing: Methods and Protocols*, Methods in Molecular Biology, vol. 2027, https://doi.org/10.1007/978-1-4939-9616-2_5, © Springer Science+Business Media, LLC, part of Springer Nature 2019

presented with the addition of a polymer host matrix. Traditional thin-film deposition techniques, such as thermal evaporation and photolithography, are not possible. For these reasons, we are detailing a method of contact printing—or "PDMS stamping"—thin films of QD/polymer to function as chemical sensors. Quantum dot layers have been contact printed to fabricate other devices including light-emitting diodes [9, 10], full-color displays [11, 12], photovoltaics [13], and optical elements [14]. In all of these applications, contact printing methods provide several advantages, including solvent-free deposition, ability to process sensors outside of a clean room, and deposition without the need for expensive photolithographic equipment.

The procedure that follows is specific for fabricating the five thin films of QD and polymer for this particular sensor array. This procedure can also serve as a template for additional materials such as metal films [15], conjugated polymers [16], and biological molecules [17], where solvent and vacuum-based deposition methods present significant technical challenges.

2 Materials

All materials are obtained at the highest purity available.

1. Polydimethyl siloxane (PDMS): Sylgard 182 silicone elastomer encapsulation kit. The two parts of this elastomer are mixed at a ratio of 10 parts elastomers to 1 part curing agent by volume.
2. Quantum dots: The solubility properties of the QDs and matrix polymers need to match to prevent phase separation in the spin coating solutions. QD and polymers that can be dispersed in toluene are used in this preparation. QDs for this method are Lumidot CdSe core dots emitting at 480 nm which are stabilized with hexadecylamine. Our laboratory has also successfully used organic soluble CdSe core dots from NN Labs (Fayetteville, AR) and QDs synthesized by using CdO as a precursor, described by Peng [18].
3. Polymers: Polymers, as received: poly(vinyl stearate) (M_w = 90,000), poly(benzyl methacrylate) (M_w = 70,000), poly(methyl methacrylate) (M_w = 15,000), poly(caprolactone) (M_w = 14,000), poly(ethylene-co-vinyl acetate), and polyisobutylene(M_w = 500,000).
4. Solvents, as received: toluene, hexane, and acetone.
5. Sensor substrates: Quartz substrates were purchased from Pyromatics (Mentor, OH). Substrates were 25 × 75 × 1 mm, with the short ends polished to an optical finish. Quartz substrates were selected to ensure optimum transmittance of the excitation light, even at 365 nm. These particular substrates

were selected because we were able to obtain them with an optical polish on the short ends, which helped with coupling of the fiber to the slide for the purpose of forming a waveguide.

6. ½ × 6 × 6 in. plate glass.
7. Lint-free cloths and swabs.
8. Razor blades.
9. Kapton tape.
10. Vacuum chamber.
11. Oven.
12. Scale.
13. Glass Pasteur pipettes.
14. Compressed N_2.
15. Spin coater with vacuum chuck.
16. Hot plate.
17. Circular hole punch (diameter 6 mm).
18. Tweezers.
19. Flow channel and sensor block: Poly(tetrafluoroethylene) (PTFE) sheeting, custom-built sensor block (*see* Subheading 3.5.2).
20. Fluorescence signal monitoring: SMA fiber-coupled 3.5 mW 365 LED, fiber-optic bundle that can be arranged linearly, digital camera with long-pass filter.
21. Data collection, post-processing, data analysis, and plotting were performed using Python 2.7 with scipy, numpy, pandas, scikit-learn, and matplotlib modules. Point Grey FlyCapture2 SDK to communicate with the camera. Similar methods could be performed with Matlab (*see* **Note 1**).

3 Methods

3.1 PDMS Substrates

1. Fabricate a mold to cure the PDMS on ½ × 6 × 6 in. thick plate glass. Clean the glass with detergent and water, and then rinse with acetone. Any physical contact should be made with a lint-free cloth.
2. Using a razor blade, remove the center of a ¼ × 6 × 6 in. piece of silicone rubber, leaving behind a border which is roughly ¼ in. wide. Secure the resulting gasket to the glass with kapton tape on the outer edge, creating a retaining wall around the perimeter of the plate glass.
3. PDMS stamps are made using a Sylgard 182 kit. Thoroughly mix two parts of the elastomer in a 10–1 ratio by weight in an aluminum weighing dish. The volume of the elastomer used is

based on the area of the mold and desired thickness. Stamps of ~1 mm thick are typical in our mold when using 20 mL of elastomer.

4. Pour the uncured PDMS into the mold to the desired thickness, and place the mold in a vacuum chamber to degas for ~30 min (*see* **Note 2**).
5. After the appearance of bubbles has stopped, transfer the mold to an oven and cure at 120 °C for 1 h.
6. Remove the mold from the oven and allow to cool to room temperature. From the large PDMS slab, cut small slabs of ~5 × 5 cm using a razor blade, avoiding the edges of the mold where the PDMS may not be flat due to uncured resin adhering to the gasket.

3.2 Quantum Dot and Polymer Solutions

To illustrate the preparation of a typical polymer and quantum dot solution, the following example describes the preparation of a poly (vinyl stearate) and Lumidot CdSe QD. While this preparation relates to a specific set of compounds, a different polymer or QD could easily be substituted to make a wide variety of solutions.

1. To produce 5 mL of a solution comprised of 20 mg/mL poly (vinyl stearate) and 13.3 mg/mL QD, dispense a 6.65 mL sample of the QD solution (as purchased) into a vial (*see* **Notes 3–8**). Using low heat (<50 °C), evaporate the solution to near dryness.
2. Make a stock solution of 50 mg/mL of poly(vinyl stearate) by weighing out 1000 mg of polymer and dissolving it in 20 mL of toluene (*see* **Note 8**).
3. Transfer the reduced QD solution to a 5 mL volumetric flask with a glass Pasteur pipette.
4. Add 2 mL of poly(vinyl stearate) stock solution to the 5 mL volumetric flask.
5. Dilute the solution in the volumetric flask with pure toluene to a volume of 5 mL.

3.3 QD/Polymer Stamp Preparation

1. Place a slab of PDMS (~5 × 5 cm) into a jar of hexane and soak for 12 h to remove any unreacted monomer (*see* **Note 9**). The slabs swell with hexane during the soak, so prior to their use place the stamps back into the vacuum chamber until they return to their original shape.
2. Place the soaked and dried 5 × 5 cm PDMS slab on a 3 in. diameter silicon wafer that has been blown free of particle contaminates with compressed N_2. Lay the PDMS on the wafer, taking care to do so in a manner that does not trap air bubbles between the silicon wafer and the PDMS, as trapping

bubbles below a thin surface will adversely affect the surface topography, and therefore the uniformity of the stamp.

3. Place the silicon wafer with PDMS on the vacuum chuck of the spin coater, and then wet the surface of the PDMS with toluene using a lint-free swab (*see* **Note 10**).
4. Using a glass Pasteur pipet, deposit ~0.5 mL of the polymer/QD solution in the center of the PDMS.
5. Using the tip of the pipet, break the surface tension of the liquid, allowing the liquid to spread across the PDMS while avoiding touching the PDMS surface with the pipet tip (*see* **Note 10**).
6. The spin coater is then engaged. Start the program with a 3-s spread spin at 500 rpm, followed by a spin at 3000 rpm for 30 s (*see* **Notes 11** and **12**).
7. Remove the silicon wafer from the spin coater and place on a hot plate to dry the QD/polymer layer for roughly 10 min. The temperature of the hot plate and the time required depend on the solvent used in the preparation of the stamp. For instance, if the solvent is toluene, the hot plate is usually kept at 80 °C, while a solution made using chloroform would be kept at 45–50 °C.
8. Once dry, remove the PDMS from the Si, place in a plastic petri dish, then wrap in Al foil, and store at 2–8 °C until needed.

3.4 Contact Printing of Sensor Pixels

1. Using a circular punch with a diameter of 6 mm, cut stamps from the PDMS. Special attention should be given to avoid any areas where there is nonuniformity of the QD/polymer layer.
2. Using tweezers and handling the stamps by the edges, place the circles of PDMS QD/polymer side down on the quartz substrates.
3. Place the quartz substrate on a 130 °C hot plate for 30 s (*see* **Note 13**).
4. Remove the quartz substrates from the hot plate and place on the lab bench to cool slightly so that they can be manipulated by hand.
5. Working with tweezers, and grabbing the edge of the PDMS, lift the stamp off the quartz substrate, working from one side to the other in one smooth motion.
6. Repeat **steps 1–5** with the desired polymer/QD stamp to produce each sensor pixel in the array.

3.5 Sensor Hardware

To produce controlled exposures of analytes to the sensor array, a sampling system is constructed that encapsulates the sensor pixels in a flow channel (*see* **Notes 14** and **15**).

3.5.1 Flow Channel

1. Cut two pieces of PTFE sheeting of 0.5 mm thickness to the same dimensions as the sensor substrate, 25 × 75 mm.
2. To define the flow channel length, cut two holes into the first sheet of PTFE, 4 mm wide and centered on the long axis of the sheet, with a distance between the two long enough to span all of the contact-printed sensor pixels.
3. On the second sheet, cut a 4 mm wide slit connecting the two holes. A schematic of the two pieces is in Fig. 1.
4. To create the flow channel, stack the PTFE sheets on the sensor substrate, as shown in Fig. 2. From the bottom of the stack are the sensor substrate, PTFE with slit, and PTFE with holes.
5. Load the stack into a custom-fabricated block (*see* next section) that holds it together and keeps the backside of the substrate clear for monitoring the fluorescent signal. The sensor block

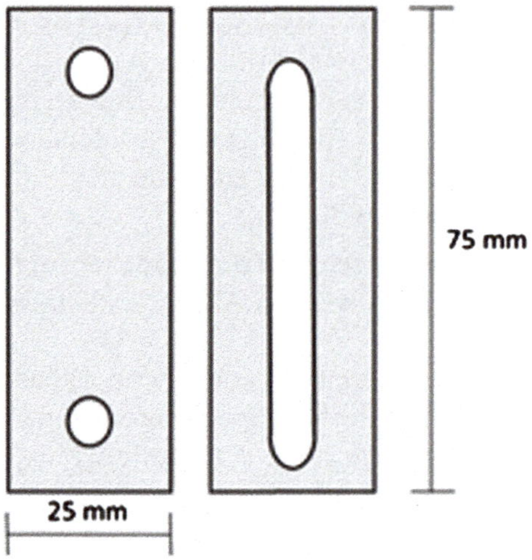

Fig. 1 Schematic representation of the two pieces of the flow channel gasket

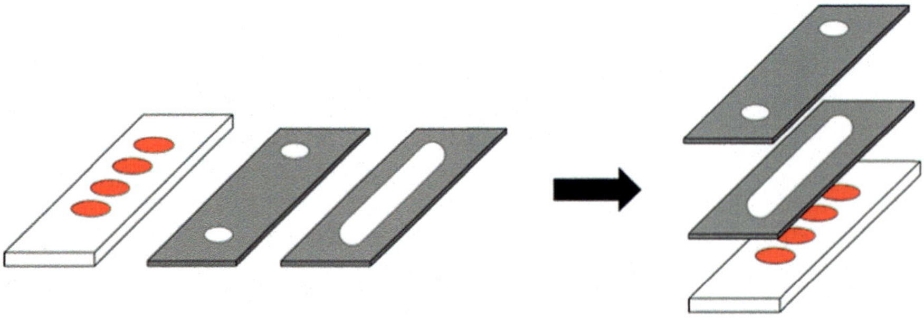

Fig. 2 Schematic representation of the two-piece flow channel and the sensor patches on a quartz slide. The two-piece flow channel is used to make a flow channel, which is 4 × 60 × 0.4 mm along the sensor patches

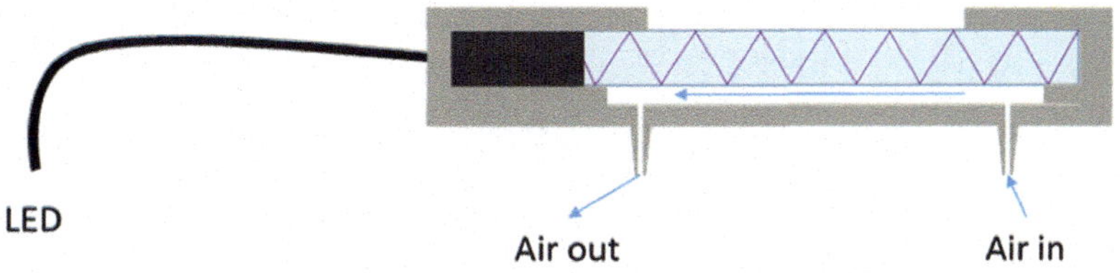

Fig. 3 Profile representation of the sensor block, with a fiber-optic cable coupled to the quartz slide. Not shown in this image is the camera used to observe the sensor from above

has a threaded port over the hole in the gasket, which allows an airtight connection to be made.

6. Using the airtight connection, pull a vacuum through a mass flow controller on one of the ports, creating flow through the channel and over the sensor pixels.

3.5.2 Sensor Block

1. A custom sensor block was constructed that holds together the substrate, the two Teflon layers which comprise the flow channel, the illumination source, and the output to the mass flow controller as shown in Fig. 3.
2. This sensor block is a two-piece system that allows for the quick and consistent mounting of the sample substrate. The block used in this study was milled from aluminum, but it could easily be 3D printed from acrylonitrile butadiene styrene (ABS) or poly(lactic acid) (PLA), provided that the application does not include materials which would degrade the polymer.
3. The sensor block allows the camera to be rigidly mounted in a position that enables clear viewing of the stamped sensor patches. Ensuring that the camera and sensor substrate are rigidly connected reduces error by ensuring that the sensor intensity readings do not change due to misalignment or changes in focus.

3.5.3 Fluorescence Signal Monitoring

1. The fluorescence excitation source is a 3.5 mW 365 LED which is SMA fiber coupled.
2. The fiber optic connecting the LED and the sensor array is a bundle of fibers on one end and a linear array on the other, as shown in Fig. 4.
3. In the sensor housing, position the linear array of the fibers to align with the polished end of the quartz substrate. This produces an evanescent wave guide, which excites the sensor material stamped on top of the substrate.
4. Monitor the fluorescent emissions from the contact-printed sensor pixels with a digital camera.

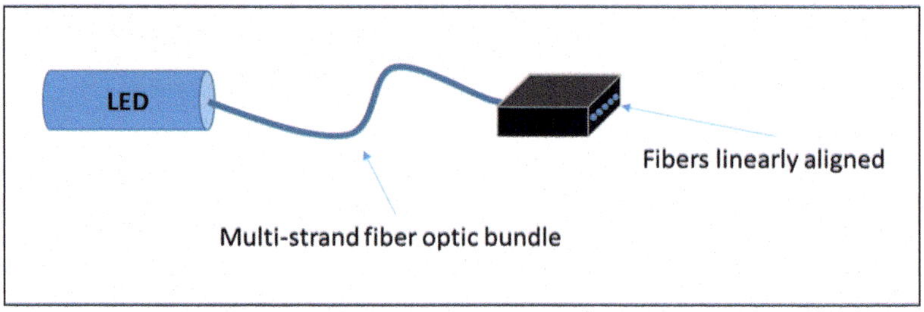

Fig. 4 Cartoon of light source, connected to a fiber-optic bundle which is realigned into a linear array in order to couple with the polished end of the quartz substrate

5. The portion of each QD/polymer sensor pixel which is located within the flow channel is indicated as a region of interest (ROI) in the collection software.
6. Monitor the average intensity of each ROI throughout each experiment, and record data at 15 Hz.

3.6 Data Processing and Feature Extraction

1. The data generated by the sensor array should be five traces of time versus intensity (*see* **Note 1**).
2. Apply a 20-point moving average applied to each data stream to smooth the data.
3. Remove sensor drift using a first-order linear fit calculation.
4. The feature vector is a set of response parameters including max change, and area of the response curve.
5. Fit third-degree polynomials to the front and back of the response curve and insert the coefficients into the feature vector [19].

3.7 Classification Methods

1. Using the feature vector created to describe the sensor response, the response can be visualized using principal component analysis (PCA) (*see* **Note 1**).
2. Using the feature vector, the identification ability of the array can be determined using linear discriminant analysis (LDA).

4 Notes

1. A Python script was used to read a user-generated input CSV file specifying the parameters for each measurement. Parameters include baseline time, sample time, recovery time, voltage of the excitation source, and sample flow rate. In this manner, many repetitions can be performed in a standardized and reproducible manner, generating large volumes of repeatable data. The script also allows the user to view camera output

in real time and define regions of interest (ROIs) to be recorded over the course of the experiment. Data from each ROI was streamed using the Point Grey FlyCapture2 SDK and recorded for later analysis.

The average intensity of the camera pixels inside each ROI was recorded at 15 Hz, giving the raw data stream of intensity versus time at each ROI, corresponding to each of the QD/polymer composites. The raw data was divided into three sections (baseline, sampling, and recovery) capturing the sensor response in the presence of background atmosphere and during a brief exposure to the analyte. The sensor was tested in a background of laboratory air with the sample pulses being introduced by the autosampler, as described in **Notes 14** and **15**. During preprocessing, the data was smoothed using a moving window average, with a window size of 20 points. Additionally, a drift correction was applied to the data based on any drift observed in the baseline data. It was assumed that any data collected during the sampling and recovery periods was dominated by chemical adsorption/desorption kinetics, and thus was not considered in the drift correction.

The drift-corrected fluorescence intensity versus time data was used to generate a third-degree polynomial expression. The coefficients from this polynomial expression were used to develop the feature vector used in a principal component analysis (PCA) of the data. Analysis of the sensor data when exposed to laboratory solvents using PCA shows that the data is highly dimensional requiring four dimensions to account for 95% of the variance.

2. When mixing PDMS, it is important to mix the two components very thoroughly. In the process of stirring, considerable quantities of air can become trapped in the viscous liquid. The presence of air bubbles in the final PDMS slabs is undesirable, as the bubbles tend to form surface imperfectionism, which disrupts the uniformity of the thin film created on top of the PDMS surface. In order to remove the air bubbles, the PDMS is placed in a vacuum chamber at room temperature. At elevated temperatures, the PDMS does become less viscous which would speed degassing; however, elevated temperature causes the curing process to proceed more quickly and can lead to the entrapment of bubbles. Furthermore, with sufficient heat and vacuum, the curing agent can begin to evaporate, trapping vapor within the slab. For this reason, it is recommended that degassing occur at room temperature and the heating process be completed without applied vacuum.

3. In this study, polymer stamping solutions were made as a comparison to inkjet print solutions. For this reason, polymer concentrations were kept between 2.5 and 20 mg/mL of the

desired polymer. Stamping solutions up to 50 mg/mL have been found to be effective in other studies.

4. Solutions for contact printing stamps in this study were produced using five different polymers: poly(vinyl stearate), poly (benzyl methacrylate), poly(methyl methacrylate), poly(caprolactone), and poly(ethylene-co-vinyl acetate).
5. Solutions were produced by dissolving the selected polymer in toluene to match the solvent in the as-received QD solutions. Other studies have found a wide variety of solvents to be useful, including, but not limited to, chloroform, chlorobenzene, THF, and hexane.
6. The solutions used to make the stamps were a mixture of polymers and QDs. All stamping solutions in this study had a polymer-to-QD ratio which was fixed at 3.3–5 mg, respectively. Other studies have shown polymer-to-QD ratios as high as 10 mg to 1 mg, respectively, and as low as 1 mg to 1 mg, respectively, to be effective.
7. The QD used in the study was Lumidot© CdSe core nanocrystals. A variety of emission wavelengths from 480 to 640 nm were used. The core QD used in this study was selected specifically to maximize reactivity with analytes. The drawback to this is that the QD is also more reactive with environmental factors such as air and water vapor, which degrade the QDs and reduce their life span. To mitigate these effects, the QD should be kept refrigerated and should be kept away from light.
8. In a sealed vessel at room temperature, solutions were prepared and allowed to stand overnight. In order to ensure a smooth and continuous layer after spinning, the solution was filtered through a 0.2 μm syringe filter prior to spinning.
9. PDMS slabs often contain small amounts of unreacted monomer after preparation. This unreacted monomer can concentrate on the surface of the PDMS slab, which interferes with the wettability of the surface, causing the spin solution to form a nonuniform coating. In order to remove the unreacted monomer, the slabs are placed into a solvent bath (typically hexane) overnight. While in the solvent bath, the slabs will swell considerably. After thoroughly soaking in solvent, the slabs are placed into a vacuum chamber to remove the solvent. A visual indicator that all solvent has been removed is that the edges of the slabs will no longer curl, but will lie flat.
10. Most solvents used will bead up on the surface of the stamp, rather than wetting the surface of the PDMS. Upon spinning, the solution will not properly adhere to surfaces which have not been wetted by the solvent. In order to ensure wetting of the surface, the tip of the pipette used to deposit the sample was used to spread the bead across the PDMS surface.

11. Spin speed, ramp rate, and duration varied broadly based on the characteristics of the solution. Typical spin profiles included a low-speed spin which allowed the solution to spread out, and a high-speed spin which removed the excess material and allowed the solution to dry. The spin parameters can be adjusted to control the thickness of the stamp.
12. The spin coater used was the Laurel Technologies WS-650. Typical spin speeds were 500–3000 rpm. Typical ramp rates were 800–1200 rpm/s. The initial, low-speed spin was 500–1000 rpm for 5–10 s. The high-speed spin was typically in the range of 2500–3000 rpm and lasted for 30–60 s. The details of the spin speed and timing at each speed are heavily influenced by the concentration of the solution and the solvent used.
13. Transfer of the stamp from the PDMS to the substrate is affected by the glass transition temperature (T_g) of the polymer being transferred. When the polymer is in a glassy state, it is more likely to crack or break up and therefore less likely to transfer to the substrate as a continuous film. In order to produce a uniform sample, therefore, the PDMS stamp is placed on the substrate and placed on a warmed hot plate to heat the stamp. Once heated, the PDMS is removed from the substrate. For polymers with a high T_g, such as poly(methyl methacrylate) (T_g = 125 °C), the hot plate was heated to 130 °C. The temperature of the hot plate will vary as a function of the T_g of the polymer being stamped. Due to a small sample size, the polymer is able to reach its T_g in 20–30 s. Under careful observation, it is possible to see a "shimmer" in the polymer as the index of refraction of the polymer changes. As soon as this "shimmer" is seen, the sample can be removed from heat and allowed to cool. The warm sample will be more pliable than when it is in its glassy state, allowing it to make conformal contact with the substrate. As it cools, the sample will preferentially adhere to the quartz substrate due to the higher surface energy. At this point the PDMS stamp can be removed, leaving the polymer on the substrate. Some of the polymers used in this study, including poly(caprolactone) and poly(ethylene-co-vinyl acetate), have a T_g which is well below room temperature. For these materials, heating the substrate is not necessary, as the sample is already above its T_g and able to make conformal contact with the substrate; however, if transfer of the sample is difficult, gentle warming often helps.
14. The sensor array was mounted into a sampling head and the polished edge of the quartz waveguide was illuminated with the 365 nm LED. The digital camera was mounted normal to the sensor surface so that it could observe all sensor pixels throughout the duration of a sampling run. A long-pass filter

was used to ensure that light from the illuminating LED did not interfere with the measurements. The software enabled the user to observe the sensor through the camera and define a region of interest (ROI) across each sensor pixel. Throughout the duration of a sampling run, the light intensity at each ROI was monitored as a function of time.

15. The autosampler provided a constant flow of air to the sensor via a mass flow controller during each test, keeping the same flow rate whether the sensor was exposed to a sample or to background air. This system allowed large amounts of data to be collected in a standardized and consistent manner.

References

1. Wilson AD, Baietto M (2011) Advances in electronic-nose technologies developed for biomedical applications. Sensors 11:1105–1176. https://doi.org/10.3390/s110101105
2. Arshak K, Moore E, Lyons GM, Harris J, Clifford S (2004) A review of gas sensors employed in electronic nose applications. Sensor Rev 24 (2):181–198. https://doi.org/10.1108/02602280410525977
3. Bright CJ, Nallon EC, Polcha MP, Schnee V (2015) Quantum dot and polymer composite cross-reactive array for chemical vapor detection. Anal Chem 87(24):12270–12275. https://doi.org/10.1021/acs.analchem.5b03559
4. Schnee VP, Bright CJ, Nallon EC, Polcha MP (2016) Contact printing of a quantum dot and polymer cross-reactive array sensor. Sens Actuat B Chem 236:506–511. https://doi.org/10.1016/j.snb.2016.06.036
5. Peveler WJ, Roldan A, Hollingsworth N, Porter MJ, Parkin IP (2016) Multichannel detection and differentiation of explosives with a quantum dot array. ACS Nano 10:1139–1146. https://doi.org/10.1021/acsnano.5b06433
6. Carlson A, Bowen AM, Huang Y, Nuzzo RG, Rogers JA (2012) Transfer printing techniques for materials assembly and micro/nanodevice fabrication. Adv Mater 24:5284–5318. https://doi.org/10.1002/adma.201201386
7. Bodas D, Khan-Malek C (2007) Direct patterning of quantum dots on structured PDMS surface. Sens Actuat B Chem 128:168–172. https://doi.org/10.1016/j.snb.2007.05.043
8. Hampton MJ, Templeton TL, DeSimone JM (2010) Direct patterning of CdSe quantum dots into sub-100 nm structures. Langmuir 26:3012–3015. https://doi.org/10.1021/la904787k
9. Kim L, Anikeeva PO, Coe-Sullivan SA, Steckel JS, Bawendi MG, Bulović V (2008) Contact printing of quantum dot light-emitting devices. Nano Lett 8:4513–4517. https://doi.org/10.1021/nl8025218
10. Anikeeva PO, Halpert JE, Bawendi MG, Bulović V (2009) Quantum dot light-emitting devices with electroluminescence tunable over the entire visible spectrum. Nano Lett 9:2532–2536. https://doi.org/10.1021/nl9002969
11. Wood V, Panzer MJ, Chen J, Bradley MS, Halpert JE, Bawendi MG, Bulović V (2009) Inkjet-printed quantum dot-polymer composites for full-color AC-driven displays. Adv Mater 21:2151–2155. https://doi.org/10.1002/adma.200803256
12. Kim TH, Cho KS, Lee EK, Lee SJ, Chae J, Kim JW et al (2011) Full-colour quantum dot displays fabricated by transfer printing. Nat Photonics 5:176–182. https://doi.org/10.1038/nphoton.2011.12
13. Arango AC, Oertel DC, Xu Y, Bawendi MG, Bulović V (2009) Heterojunction photovoltaics using printed colloidal quantum dots as a photosensitive layer. Nano Lett 9(2):860–863. https://doi.org/10.1021/nl803760j
14. Shallcross RC, Chawla GS, Marikkar FS, Tolbert S, Pyun J, Armstrong NR (2009) Efficient CdSe nanocrystal diffraction gratings prepared by microcontact molding. ACS Nano 3(11):3629–3637. https://doi.org/10.1021/nn900735y
15. Schmid H, Wolf H, Allenspach R, Riel H, Karg S, Michel B, Delamarche E (2003) Preparation of metallic films on elastomeric stamps and their application for contact processing and contact printing. Adv Funct Mater

13:145–153. https://doi.org/10.1002/adfm.200390021

16. Yim KH, Zheng Z, Liang Z, Friend RH, Huck WTS, Kim JS (2008) Efficient conjugated-polymer optoelectronic devices fabricated by thin-film transfer-printing technique. Adv Funct Mater 18:1012–1019. https://doi.org/10.1002/adfm.200701321
17. Mayer M, Yang J, Gitlin I, Gracias DH, Whitesides GM (2004) Micropatterned agarose gels for stamping arrays of proteins and gradients of proteins. Proteomics 4(8):2366–2376. https://doi.org/10.1002/pmic.200300748
18. Peng ZA, Peng X (2001) Formation of high-quality CdTe, CdSe, and CdS nanocrystals using CdO as precursor. J Am Chem Soc 123 (1):183–184. https://doi.org/10.1021/ja003633m
19. Nallon EC, Schnee VP, Bright C, Polcha MP, Li Q (2015) Discrimination enhancement with transient feature analysis of a graphene chemical sensor. Anal Chem 88(2):1401–1406. https://doi.org/10.1021/acs.analchem.5b04050

Chapter 6

Colorimetric Sensor Array Based on Amino Acid-Modified Gold Nanoparticles for Toxic Metal Ion Detection in Water

Gülsu Şener and Adil Denizli

Abstract

Several chromatographic and spectroscopic methods are available for the detection of toxic mercury (Hg^{2+}) in water; however simple, rapid, inexpensive, and sensitive methods are still needed. In this chapter, we describe a facile, very sensitive, and rapid method for the colorimetric detection of Hg^{2+} in water with the detection limit of 2.9 nM. This simple procedure is based on the lysine-induced aggregation of citrate-capped gold nanoparticles (AuNPs) in the presence of Hg^{2+} ions.

Key words Gold nanoparticles, Colorimetric sensors, Mercury ion detection, Label free, Lysine

1 Introduction

Monitoring of water contamination by toxic metal ions is very important for improving public health as some heavy metal ions such as mercury, lead, and cadmium can cause serious health and environmental effects [1–4]. Several highly sensitive and selective protocols are available for detection of heavy metal ion contaminations in water based on chromatographic and spectroscopic techniques such as high-performance liquid chromatography (HPLC), inductively coupled plasma mass spectrometry (ICP-MS), and atomic absorption spectrometry (AAS) [5–8]. However, these methods require expensive instruments and they are time consuming and labor extensive. Therefore, there is a need for development of simple but sensitive methods for rapid and reliable analysis of heavy metal ion levels in water resources.

In this chapter we describe a method for sensitive and selective detection of solvated divalent mercuric ion (Hg^{2+}), which causes great toxicity mainly on renal and nervous systems through the disruption of enzyme activity [9, 10]. This facile method utilizes as-prepared citrate-capped gold nanoparticles (AuNPs) and an amino acid, lysine, to induce the aggregation of AuNPs in the

Jessica E. Fitzgerald and Hicham Fenniri (eds.), *Biomimetic Sensing: Methods and Protocols*, Methods in Molecular Biology, vol. 2027, https://doi.org/10.1007/978-1-4939-9616-2_6,

presence of low concentrations of Hg^{2+} ions. Aggregation of AuNPs results in a color change of the AuNP dispersion from red to purple or gray. By analysis of this color change using a simple UV–Vis absorption spectrophotometer, sensitive detection of Hg^{2+} ions (limit of detection is 2.9 nM) can be achieved within 5 min from real-world samples and in the presence of other metal ions [11].

2 Materials

Prepare all solutions using ultrapure water (18 MΩ-cm at 25 °C) and analytical grade reagents. Prepare and store all reagents at room temperature (unless indicated otherwise) (*see* **Note 1**).

1. Stock sodium citrate solution: Dissolve 0.5 g of sodium citrate tribasic dehydrate in 50 mL water.
2. Gold(III) chloride trihydrate solution: Weigh 8.5 mg gold(III) chloride trihydrate and dissolve in 5 mL of water (*see* **Note 2**).
3. Metal ion stock solutions: Prepare all metal ion solutions from nitrate salts to give the molarity of 0.3 mM (50 mL). Except sodium, to prepare Na ion solution use NaCl. For other metal ion solutions use the following salts: $Cu(NO_3)_2{\cdot}6H_2O$, $Ni(NO_3)_2{\cdot}6H_2O$, $Pb(NO_3)_2$, $Fe(NO_3)_3{\cdot}9H_2O$, $Hg(NO_3)_2{\cdot}H_2O$, $AgNO_3$, NaCl, $Zn(NO_3)_2{\cdot}6H_2O$, $Co(NO_3)_2{\cdot}6H_2O$, KNO_3, $Ca(NO_3)_2{\cdot}4H_2O$, $Al(NO_3)_3{\cdot}H_2O$, and $Cd(NO_3)_2{\cdot}4H_2O$.
4. Amino acid stock solutions: Dissolve appropriate amounts of the following amino acids in 10 mL of water to give final amino acid concentration of 1 mM: lysine, cysteine, histidine, tyrosine, aspartic acid, arginine, alanine, glycine, phenylalanine, tryptophan, methionine, leucine, and isoleucine.
5. Flasks.
6. Magnetic stir plate (with heater).
7. UV–Vis spectrometer.
8. Cuvettes.
9. Pipettes.
10. Scale.

3 Methods

3.1 Synthesis of Citrate-Capped Gold Nanoparticles

The AuNPs were prepared according to Turkevich method in which $HAuCl_4$ salt is reduced to Au using sodium citrate [12].

1. Secure a 250 mL flask over the center of a magnetic stir plate.
2. Heat 12 mL of sodium citrate solution (1%, w/w) close to boiling (*see* **Note 3**).

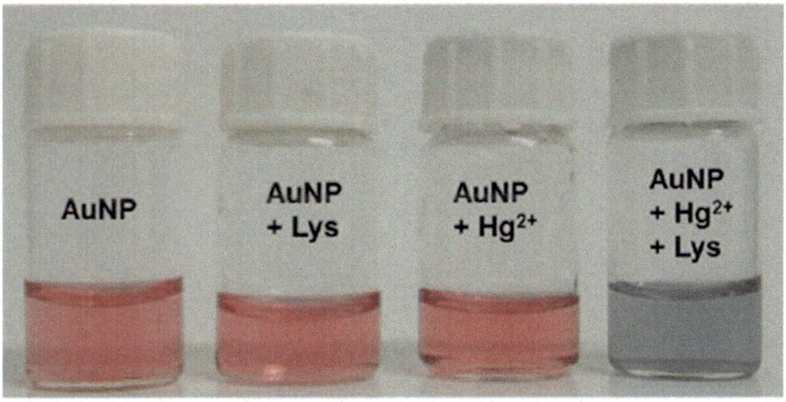

Fig. 1 Photographs of AuNP solutions in the presence of lysine (0.4 mM) and Hg^{2+} (10 μM)

3. Pour 95 mL of water into the flask, stir the water at 300 rpm using a magnetic stir bar, and heat the water to the boiling state.
4. Add 5 ml of gold(III) chloride trihydrate solution to the boiling water.
5. Add preheated sodium citrate solution to the above boiling solution.
6. Stir the solution vigorously while heating (keep at boiling state) for 20 min until color of the solution turns to deep red.
7. Let the solution to be cooled to room temperature.
8. Bring the volume of the AuNP solution to 100 mL by adding water.
9. Final concentration of AuNP in this solution was 0.4 nM, which was calculated according to a previous report [13].
10. Characterize the as prepared by using an UV–Vis spectrophotometer. The AuNPs prepared according to this protocol appear red with an extinction maximum around 525 nm (Fig. 1).

3.2 Colorimetric Sensing of Hg^{2+}

1. Mix 0.5 mL of AuNP solution with 0.2 mL of Hg^{2+} solutions at different concentrations to give final Hg^{2+} concentrations in the range of 1 nM–20 μM. Add 0.2 mL water as control (blank samples).
2. Add 0.5 mL of lysine solution to this solution and well mix the solution.
3. Incubate this solution at room temperature for 5 min at static conditions.
4. Detect the color changes by naked eye and/or record the UV–Vis absorption spectrum.

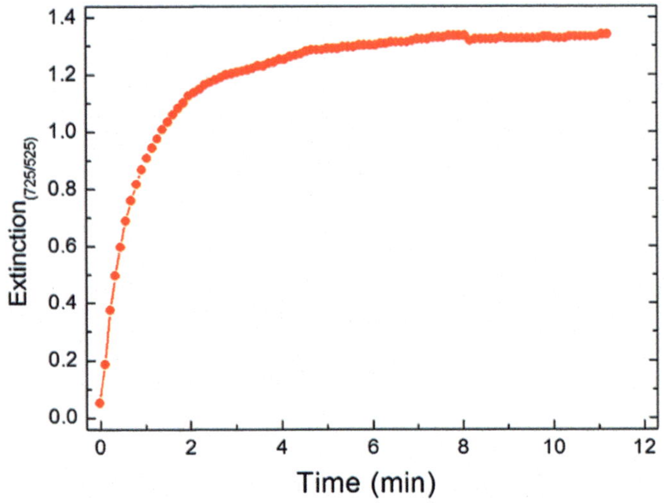

Fig. 2 Fast response of the colorimetric assay against increasing Hg^{2+} concentrations, which is obtained by taking the ratio of the extinction of AuNP suspension at 725 nm to the extinction at 525 nm

5. Perform the absorption measurements using a UV–Vis spectrophotometer between 350 nm and 850 nm.
6. Calculate the sensor responses by dividing the extinction of AuNP suspension at 725 nm to the extinction at 525 nm (Ex_{725}/Ex_{525}) (*see* **Note 4**) (Fig. 2).

3.3 Recovery Experiments from Tap Water

1. Mix 0.5 mL of AuNP solution with 0.2 mL of tap water samples spiked with Hg^{2+} solutions to give final Hg^{2+} concentrations in the range of 1 nM–20 μM.
2. Add lysine solutions and detect color changes as described in Subheading 3.2.
3. Use the response (Ex_{725}/Ex_{525}) of the colorimetric assay against spiked tap water samples and plot the calibration (Fig. 3).
4. Use the calibration curve to calculate the Hg^{2+} concentration in the tap water samples.
5. Use Eq. 1 to calculate recovery values:

$$\text{Recovery } (\%) = (\text{calculated Hg}^{2+}/\text{added Hg}^{2+}) \times 100\% \quad (1)$$

3.4 Effect of Other Amino Acids on the Response Colorimetric Sensor

1. Mix 0.5 mL of AuNP solution with 0.2 mL of Hg^{2+} solution to give final Hg^{2+} concentration of 10 μM. Add 0.2 mL water as control.
2. Add 0.5 mL of cysteine, histidine, tyrosine, aspartic acid, arginine, alanine, glycine, phenylalanine, tryptophan, methionine, leucine, or isoleucine solution in water to this solution.
3. Detect the color changes as described in Subheading 3.2.

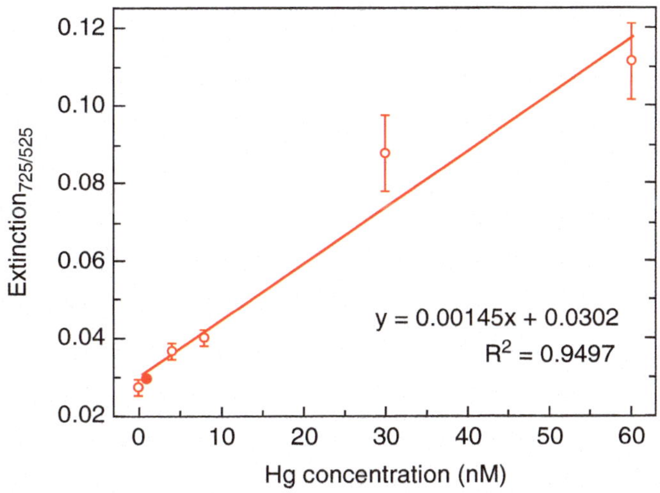

Fig. 3 Response of AuNP solutions after addition of lysine in the presence of different amounts Hg^{2+} in tap water samples

3.5 Selectivity of the Colorimetric Sensor

1. Mix 0.5 mL of AuNP solution with 0.2 mL of metal ion solutions.
2. Investigate the selectivity of the colorimetric sensor using several metal ions (Cu^{2+}, Ni^{2+}, Pb^{2+}, Co^{2+}, Zn^{2+}, Ca^{2+}, Al^{3+}, K^+, Fe^{3+}, Cd^{2+}, Ag^+, Na^+, Au^{3+}) under the abovementioned conditions.
3. Adjust the final concentration of the competing metal ions to 50 μM.
4. Add lysine solutions and detect color changes as described in Subheading 3.2.
5. Also, perform selectivity experiments in the presence of a mixture of all metal ions (each ion concentration was adjusted to 10 μM) as described above.
6. Perform control experiments in the absence of lysine.

4 Notes

1. Clean glassware and stir bars using deionized water and isopropyl alcohol.
2. Weigh gold(III) chloride trihydrate by using a Teflon spatula in order to prevent the reduction of gold ions.
3. Wear eye protection all the time. Avoid the spill of heated sodium citrate solution.
4. Preform all experiments in triplicate.

References

1. Clarkson TW, Magos L, Myers GJ (2003) The toxicology of mercury—current exposures and clinical manifestations. N Engl J Med 349:1731–1737
2. Uzun L, Kara A, Osman B, Yılmaz E, Besirli N, Denizli A (2009) Removal of heavy-metal ions by magnetic beads containing triazole chelating groups. J Appl Polym Sci 114:2246–2253
3. Chen L, Fu X, Lu W, Chen L (2013) Highly sensitive and selective colorimetric sensing of Hg^{2+} based on the morphology transition of silver nanoprisms. ACS Appl Mater Interfaces 5:284–290
4. Kim HN, Ren WX, Kim JS, Yoon J (2012) Fluorescent and colorimetric sensors for detection of lead, cadmium, and mercury ions. Chem Soc Rev 41:3210–3244
5. Hong YS, Rifkin E, Bouwer EJ (2011) Combination of diffusive gradient in a thin film probe and IC-ICP-MS for the simultaneous determination of CH_3Hg^+ and Hg^{2+} in oxic water. Environ Sci Technol 45:6429–6436
6. Erxleben H, Ruzicka J (2005) Atomic absorption spectroscopy for mercury, automated by sequential injection and miniaturized in lab-on-valve system. Anal Chem 77:5124–5128
7. Barriada JL, Tappin AD, Evans EH, Achterberg EP (2007) Dissolved silver measurements in seawater. TrAC Trends Anal Chem 26:809–817
8. Fong BMW, Siu TS, Lee JSK, Tam S (2007) Determination of mercury in whole blood and urine by inductively coupled plasma mass spectrometry. J Anal Toxicol 31:281–287
9. Kim YR, Mahajan RK, Kim JS, Kim H (2010) Highly sensitive gold nanoparticle-based colorimetric sensing of Mercury(II) through simple ligand exchange reaction in aqueous media. ACS Appl Mater Interfaces 2:292–295
10. Du J, Liu MY, Lou XH, Zhao Y, Wang Z, Xue Y, Zhao J, Xu Y (2012) Highly sensitive and selective chip-based fluorescent sensor for mercuric ion: development and comparison of turn-on and turn-off systems. Anal Chem 84:8060–8066
11. Sener G, Uzun L, Denizli A (2013) Lysine-promoted colorimetric response of gold nanoparticles: a simple assay for ultrasensitive Mercury(II) detection. Anal Chem 86:514–520
12. Enustun BV, Turkevich J (1963) Coagulation of colloidal gold. J Am Chem Soc 85:3317–3328
13. Haiss W, Thanh NTK, Aveyard J, Fernig DG (2007) Determination of size and concentration of gold nanoparticles from UV-Vis spectra. Anal Chem 79:4215–4221

Chapter 7

Identification of Several Toxic Metal Ions Using a Colorimetric Sensor Array

Gülsu Şener and Adil Denizli

Abstract

Water pollution by toxic metal ions is a worldwide environmental and health problem, and therefore monitoring of toxic metal ions in water resources is highly desired. In this chapter, we describe a simple colorimetric sensor array for simultaneous detection of multiple toxic heavy metal ions (Hg^{2+}, Cd^{2+}, Fe^{3+}, Pb^{2+}, Al^{3+}, Cu^{2+}, and Cr^{3+}) in water. This assay is produced by using 11-mercaptoundecanoic acid (MUA)-capped gold nanoparticles (AuNPs) and five amino acids (lysine, cysteine, histidine, tyrosine, and arginine). The presence of amino acids can enhance or diminish the aggregation MUA-capped AuNPs with metal ions. The color change of the sensor array after aggregation in some of the channels creates unique response patterns for each metal ion.

Key words Gold nanoparticles, Toxic metal ions, Colorimetric assays, Sensor array, Environmental monitoring

1 Introduction

As mentioned in the previous chapter, contamination of water resources by heavy metal ions can cause serious environmental and health problems [1–3]. In the previous chapter, we described a highly sensitive and selective colorimetric method for detection of one of the most toxic heavy metal ions, the solvated divalent mercuric ion (Hg^{2+}) [4]. Here, we describe a colorimetric method for detection of multiple heavy metal ions using a single colorimetric assay. While the sensitivity of this method is still far from the market requirements, the novel approach described in this chapter represents the first example (to our knowledge) of a colorimetric assay that can simultaneously and selectively detect several metal ions. This approach can be a starting point for future studies to develop colorimetric sensor arrays that can rapidly, reproducibly, selectively, and sensitively detect several heavy metal ions at the same time. That would significantly reduce the cost and time required for analysis of water samples.

Jessica E. Fitzgerald and Hicham Fenniri (eds.), *Biomimetic Sensing: Methods and Protocols*, Methods in Molecular Biology, vol. 2027, https://doi.org/10.1007/978-1-4939-9616-2_7, © Springer Science+Business Media, LLC, part of Springer Nature 2019

The colorimetric detection scheme described in this chapter is based on metal ion-induced aggregation or non-aggregation of surfactant-capped gold nanoparticles (AuNPs) in the presence of different amino acids [5]. The color of the AuNP dispersion changes upon aggregation from red to first purple and then blue [6]. The use of amino acids provides additional interactions between AuNPs and metal ions as amino acids can bind to gold surface and form complexes with metal ions through their carboxyl and amino groups or their side groups [7–10]. Therefore, metal ions, amino acids, and AuNPs can interact through several different pathways. In this method, a colorimetric sensor array is formed by using an amino acid and surfactant-stabilized AuNP couple as a distinct sensing element. Accordingly, we used lysine, cysteine, histidine, tyrosine, and arginine and mercaptoundecanoic acid-stabilized AuNPs to form five different sensing elements. Interestingly, presence of amino acids can either enhance or diminish the metal-induced aggregation, and therefore the color change. To discriminate the seven metal ions used in this study (Hg^{2+}, Cd^{2+}, Fe^{3+}, Pb^{2+}, Al^{3+}, Cu^{2+}, and Cr^{3+}) the color change in each sensing element is analyzed by using hierarchical cluster analysis or simply by naked eye.

2 Materials

Prepare all solutions using ultrapure water (18 MΩ-cm at 25 °C) and analytical grade reagents. Prepare and store all reagents at room temperature (unless indicated otherwise) (*see* **Note 1**).

1. Stock sodium citrate solution: Dissolve 0.5 g of sodium citrate tribasic dehydrate in 50 mL water.
2. Gold(III) chloride trihydrate solution: Weigh 8.5 mg gold(III) chloride trihydrate and dissolve in 5 mL of water (*see* **Note 2**).
3. MUA solution: Weigh 3.25 mg MUA and dissolve in 10 mL water containing NaOH (0.5 M, 30 μL).
4. Metal ion stock solutions: Prepare all metal ion solutions from nitrate salts to give the molarity of 0.3 mM (50 mL). Except sodium, to prepare Na ion solution use NaCl. For other metal ion solutions use the following salts: $Cu(NO_3)_2{\cdot}6H_2O$, $Ni(NO_3)_2{\cdot}6H_2O$, $Pb(NO_3)_2$, $Fe(NO_3)_3{\cdot}9H_2O$, $Hg(NO_3)_2{\cdot}H_2O$, $AgNO_3$, NaCl, $Zn(NO_3)_2{\cdot}6H_2O$, $Co(NO_3)_2{\cdot}6H_2O$, KNO_3, $Ca(NO_3)_2{\cdot}4H_2O$, $Al(NO_3)_3{\cdot}H_2O$, and $Cd(NO_3)_2{\cdot}4H_2O$.
5. Amino acid stock solutions: Dissolve appropriate amounts of the following amino acids in 10 mL of water to give final amino acid concentration of 1 mM: lysine, cysteine, histidine, tyrosine, and arginine.

6. Flasks.
7. Magnetic stir plate (with heater).
8. UV–vis spectrometer.
9. Centrifuge.
10. Cuvettes.
11. 96-Well plate
12. Pipettes.
13. Matlab software, or comparable software that can perform clustering analyses.
14. Isopropyl alcohol.
15. Teflon spatula.
16. Scale.

3 Methods

3.1 Synthesis of MUA-Capped AuNPs

The AuNPs were prepared according to Turkevich method in which $HAuCl_4$ salt is reduced to Au using sodium citrate [11].

1. Secure a 250 mL flask over the center of a magnetic stir plate.
2. Heat 12 mL sodium citrate solution (1%, w/w) (*see* **Note 3**).
3. Pour 95 mL of water into the flask, stir the water at 300 rpm using a magnetic stir bar, and heat the water to the boiling state.
4. Add 5 ml of gold(III) chloride trihydrate solution to the boiling water.
5. Add preheated sodium citrate solution to the above boiling solution.
6. Stir the solution vigorously while heating for 20 min until color of the solution turns to deep red.
7. Let the solution to be cooled to room temperature.
8. Bring the volume of the AuNP solution to 100 mL by adding water. Final concentration of AuNP in this solution was 0.1 nM, which was calculated according to a previous report [12].
9. To modify the AuNP surface with MUA, add 10 mL of MUA solution to the boiling state of citrate-capped AuNP solution (50 mL) while heating.
10. Stir the solution for 1 h [6].
11. Cool the solution to room temperature.
12. Bring the volume of the MUA-capped AuNP solution to 50 mL by adding water.

13. Wash MUA-capped AuNPs at 8092 ×*g* for 20 min with water to remove the unbound MUA molecules.
14. Repeat the washing step twice.
15. Store the MUA-capped AuNPs at 4 °C (*see* **Note 4**).

3.2 Determination of Optimum Amino Acid Concentrations in the Colorimetric Assay

1. Determine the suitable amino acid concentrations that will be used in the colorimetric sensor array by recording the UV–vis spectra of AuNPs (0.1 nM) in the presence of different amino acid concentrations between 0 μM and 500 μM at different time intervals up to 1 h.
2. Detect any aggregation by dividing the extinction of AuNPs at 625 nm by the extinction of 525 nm (Ex_{625}/Ex_{525}).
3. Determine amino acid concentrations to ensure no aggregation of AuNPs in the absence of metal ions.
4. Accordingly, amino acid concentrations of 50 μM for cysteine and 200 μM for other amino acids (lysine, histidine, tyrosine, arginine) were selected.

3.3 Colorimetric Detection of Multiple Metal Ions

1. Perform the experiments in a 96-well plate.
2. Mix 50 μL MUA-capped AuNPs (0.1 nM) with 5 μL of cysteine or 20 μL of other amino acids (lysine, cysteine, histidine, tyrosine, and arginine) at determined concentrations.
3. Add metal ions (Hg^{2+}, Cd^{2+}, Fe^{3+}, Pb^{2+}, Al^{3+}, Cu^{2+}, and Cr^{3+}) in water at different concentrations to this solution to give final metal ion concentrations in the range of 2 μM to 50 μM (Fig. 1).
4. Bring the final volume of assay to 100 μL with water.
5. Detect color changes by the naked eye and/or by using a plate reader.
6. Record the colorimetric responses at 625 nm and 525 nm.
7. Calculate the colorimetric responses by dividing the extinction of AuNPs at 625 nm by the extinction of 525 nm (Ex_{625}/Ex_{525}).

3.4 Data Analysis for Multiple Ion Detection

1. Analyze the colorimetric responses of the array against metal ions at different concentrations using hierarchical cluster analysis (HCA) as the pattern recognition technique.
2. Perform the analyses by built-in HCA algorithms of MATLAB software (*see* **Note 5**).
3. Use Euclidean distance method for obtaining the dendrograms.

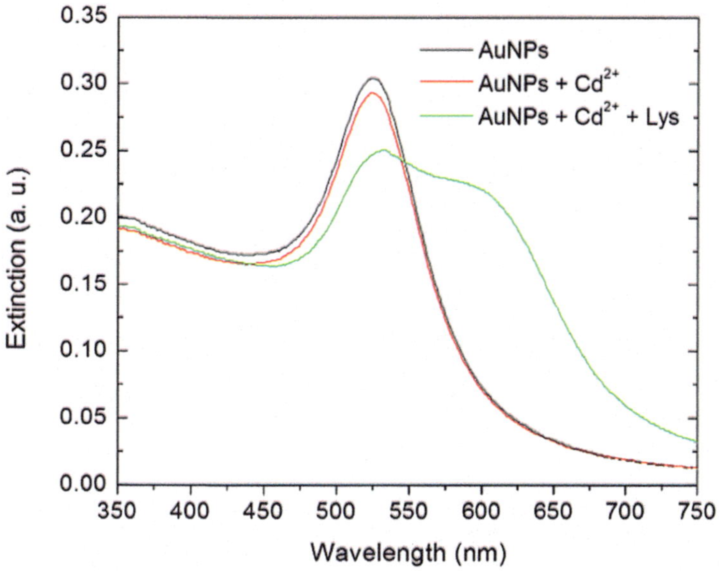

Fig. 1 Representative UV–vis spectra for AuNPs in the presence or absence of metal ions and amino acids

4 Notes

1. Clean glassware and stir bars using deionized water and isopropyl alcohol.
2. Weigh gold(III) chloride trihydrate by using a Teflon spatula in order to prevent the reduction of gold ions.
3. Wear eye protection all the time. Avoid the spill of heated sodium citrate solution.
4. MUA-capped AuNPs were stable at least for 2 months at 4 °C.
5. MATLAB code for HCA:

```
X = [enter data].
D = pdist (X).
Squareform (D).
Z = linkage (X,'complete','euclidean');
H = dendrogram (Z,'Orientation','right').
Set (H,'LineWidth',1).
```

The distance between two data points was calculated using Euclidean distance (Eq. 1):

$$d(a,b) = \sqrt{\sum_{i=1}^{n} (a_i - b_i)^2} \quad (1)$$

References

1. Clarkson TW, Magos L, Myers GJ (2003) The toxicology of mercury - Current exposures and clinical manifestations. N Engl J Med 349:1731–1737
2. Chen L, Fu X, Lu W, Chen L (2013) Highly sensitive and selective colorimetric sensing of Hg^{2+} based on the morphology transition of silver nanoprisms. ACS Appl Mater Interfaces 5:284–290
3. Kim HN, Ren WX, Kim JS, Yoon J (2012) Fluorescent and colorimetric sensors for detection of lead, cadmium, and mercury ions. Chem Soc Rev 41:3210–3244
4. Sener G, Uzun L, Denizli A (2013) Lysine-promoted colorimetric response of gold nanoparticles: a simple assay for ultrasensitive mercury(II) detection. Anal Chem 86:514–520
5. Sener G, Uzun L, Denizli A (2014) Colorimetric sensor array based on gold nanoparticles and amino acids for identification of toxic metal ions in water. ACS Appl Mater Interfaces 6:18395–18400
6. Saha K, Agasti SS, Kim C, Li X, Rotello VM (2012) Gold nanoparticles in chemical and biological sensing. Chem Rev 112:2739–2779
7. Kim Y, Johnson RC, Hupp JT (2001) Gold nanoparticle-based sensing of "Spectroscopically Silent" heavy metal ions. Nano Lett 1:165–167
8. Chai F, Wang C, Wang T, Li L, Su Z (2010) Colorimetric detection of Pb^{2+} using glutathione functionalized gold nanoparticles. ACS Appl Mater Interfaces 2:1466–1470
9. Cush R, Cronin JM, Stewart WJ, Maule CH, Molloy J, Goddard NJ (1993) The resonant mirror: a novel optical biosensor for direct sensing of biomolecular interactions. Part I: Principle of operation and associated instrumentation. Biosens Bioelectron 8:347–354
10. Promnimit S, Bera T, Baruah S, Dutta J (2011) Chitosan capped colloidal gold nanoparticles for sensing zinc ions in water. J Nano Res 16:55–61
11. Enustun BV, Turkevich J (1963) Coagulation of colloidal gold. J Am Chem Soc 85:3317–3328
12. Haiss W, Thanh NTK, Aveyard J, Fernig DG (2007) Determination of size and concentration of gold nanoparticles from UV–Vis spectra. Anal Chem 79:4215–4221

Chapter 8

Real-Time Sensing with Patterned Plasmonic Substrates and a Compact Imager Chip

Spencer T. Seiler, Isabel S. Rich, and Nathan C. Lindquist

Abstract

Optical sensing is an important research field due to its proven ability to be extremely sensitive, nondestructive, and applicable to sensing a wide range of chemical, thermal, electric, or magnetic phenomena. Beyond traditional optical sensors that often rely on bulky setups, plasmonic nanostructures can offer many advantages based on their sensitivity, compact form, cost-effectiveness, multiplexing compatibility, and compatibility with many standard semiconductor nanofabrication techniques. In particular, plasmon-enhanced optical transmission through arrays of nanostructured holes has led to the development of a new generation of optical sensors. In this chapter we present a simple fabrication technique to use plasmonic nanostructures as compact sensors. We position the nanohole array, an LED illumination source, and a spacer layer directly on top of a standard complementary metal–oxide–semiconductor (CMOS) imager chip. This setup is a viable sensor platform in both liquid and gas environments. These devices could operate as low-cost sensors for environmental monitoring, security, food safety, or monitoring small-molecule binding to extract affinity information and binding constants.

Key words Plasmonics, Lab on a chip, Optical sensors, Electronic nose

1 Introduction

The ability to accurately detect liquid- [1–4] and gas-phase [5–8] analytes is important for a wide range of areas including security, drug discovery, food safety, biomarker detection, and environmental monitoring. While many sensing technologies are available, optical sensors have significant advantages such as nondestructive remote probing, analyzing refractive index and other material properties, fingerprinting molecular vibrations, and detecting analyte concentrations to a high degree of accuracy and specificity [4, 9, 10]. But as optical devices are scaled in size to be smaller than the optical wavelength for single-molecule sensing or on-chip integration, near-field resonant devices can be used to strengthen the optical interactions with the sample. In this regard, plasmonic

Jessica E. Fitzgerald and Hicham Fenniri (eds.), *Biomimetic Sensing: Methods and Protocols*, Methods in Molecular Biology, vol. 2027, https://doi.org/10.1007/978-1-4939-9616-2_8, © Springer Science+Business Media, LLC, part of Springer Nature 2019

optical sensors [11–13] comprise an important area of research that has seen rapid development.

Plasmons are oscillations of free electrons at the surface of a metal film or metallic nanostructure. Surface plasmons that propagate over a smooth metal film extend only *ca.* 100 nm into the surrounding medium and are often used for biosensing and drug discovery [3]. Metallic nanostructures can also sustain localized surface plasmons that extend *ca.* 10 nm into the surrounding medium [14]. These have been used to detect the refractive indices of various inert gases [15] or for chemical identification via surface-enhanced Raman spectroscopy (SERS) [6, 7]. Due to these near-field effects, plasmons are extremely sensitive to surface changes and show significant potential for on-chip integration, low-cost processing, and massive sensor parallelization. For example, enhanced optical transmission through an array of plasmonic nanoholes [16] has led to the development of a new generation of optical biosensors [13, 17–22]. These nanohole array sensors have been used in multichannel assays [22], for submicron-resolution imaging [21], and with on-chip integration. Nanohole sensors are also suitable for flow-through detection [23, 24]. While many plasmonic biosensors are used in a liquid environment, vapor sensing is also an important area of research [25–30] where plasmonic nanoparticles [15, 31, 32] and nanohole arrays [30, 33, 34] have been used. To enhance the specificity of these sensors to various liquid- and gas-phase analytes, self-assembled monolayers [31, 35], electrochemical functionalization [30], and analyte-sensitive substrates [34] have shown promise.

To realize their full potential, plasmonic biosensors should be integrated and miniaturized, eliminating bulky optical setups. Imaging the transmitted intensity of plasmonic arrays directly onto a CMOS camera chip is one potential solution [36], but since plasmonic biosensors typically rely on small shifts in the resonance wavelength, important spectral information is lost. Surface plasmon imaging with nanohole arrays can be used for multiplex sensing and integration with microfluidics [21] but is often accomplished with monochromatic light, again losing important spectral information. A benchtop imaging spectrometer can capture a spectrum from several channels simultaneously but is restricted to a one-dimensional array of sensing devices [22] and requires large optical components. Therefore, several important challenges remain in the development of miniaturized, low-cost plasmonic sensors that can be integrated with an optical excitation and detection scheme as well as any relevant fluidic or vapor delivery systems.

To address some of these issues, in this chapter we demonstrate a simple plasmonic nanohole sensing system based on directly imaging a diffracted transmission spectrum with a bare CMOS imager chip and LED illumination [33]. The plasmonic nanoholes

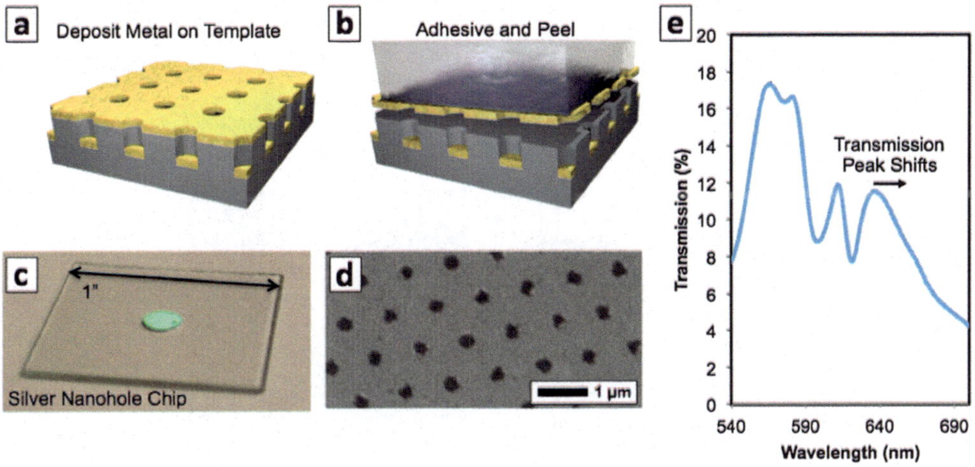

Fig. 1 Nanohole sensor fabrication. (**a**) A thin gold or silver film is deposited on the silicon template. Due to non-conformal deposition, holes are left in the upper film. (**b**) A thin layer of adhesive is used to peel the patterned film. (**c**) Large-area patterns are possible. (**d**) SEM image of a representative nanohole sample. (**e**) The optical transmission spectrum taken with a traditional fiber-optic spectrometer. During sensing experiments, the peaks will shift due to small changes in refractive index in and around the nanoholes. Adapted from Ref. 33 with permission from Institute of Physics (IOP) publishing

are shown in Fig. 1. By attaching a plasmonic nanohole array directly to the imager chip with an index-matching layer, light diffracted from the nanohole grating lands directly onto the camera pixels, thereby resolving a full spectrum from multiple sensing spots simultaneously and in real time. The CMOS camera is attached to a computer via a USB cable for data capture and analysis. By illuminating the sensing chip with an LED, this arrangement eliminates the need for an external laser, microscope, and spectrometer, dramatically simplifying the traditional nanohole-array sensing platform. This setup is shown in Fig. 2. We show in Fig. 3 that such a system can be used in both a liquid and a vapor sensing capacity. While different liquids can have a large refractive index contrast, to enhance the sensitivity of our device to ethanol vapors, we coated the chips with Reichardt's dye, a well-known solvatochromic dye [37]. The dye changes color depending on the presence and polarity of solvents and has been used previously to develop various liquid [38] or vapor [39] sensors. Finally, we demonstrate the potential for multiplexing by measuring simultaneous signals from two side-by-side locations on the CMOS imager. The development of this general-purpose, nanohole-on-imager system shows potential for low-cost, multiplexed, label-free plasmonic sensing.

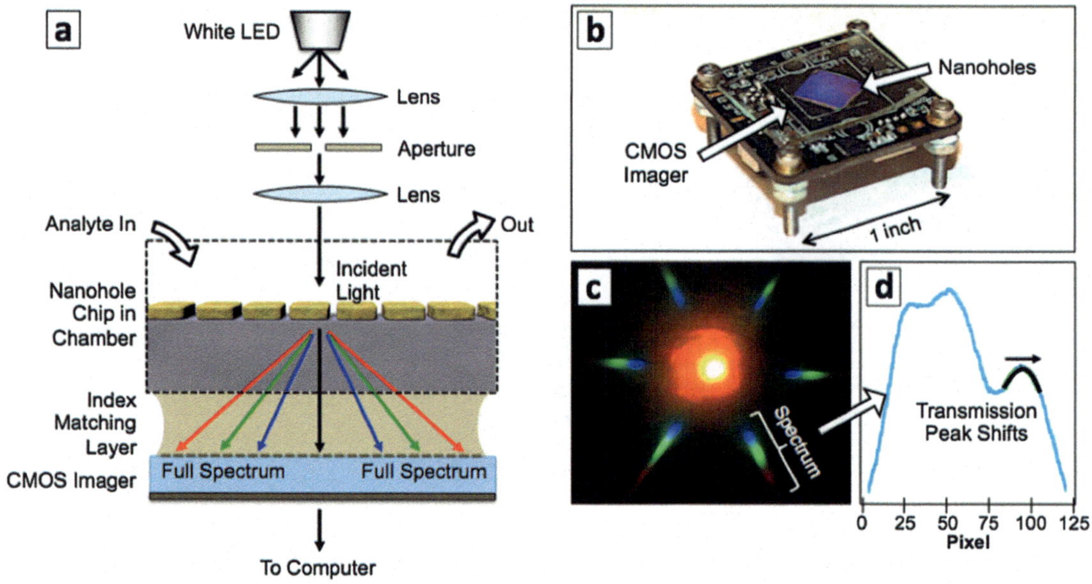

Fig. 2 Integrated nanohole sensing chip. (**a**) A white LED, simple optics, and nanohole chip are all mounted directly above a bare CMOS imager chip. Transmitted light diffracts from the nanohole array, spreading into a complete spectrum that can be imaged and used for sensing. (**b**) Photograph of the imager plus nanoholes. (**c**) Raw image from a color CMOS chip showing the diffracted spectrum. (**d**) A cut through the spectrum shows peaks that can be used for sensing. Adapted from Ref. 33 with permission from Institute of Physics (IOP) publishing

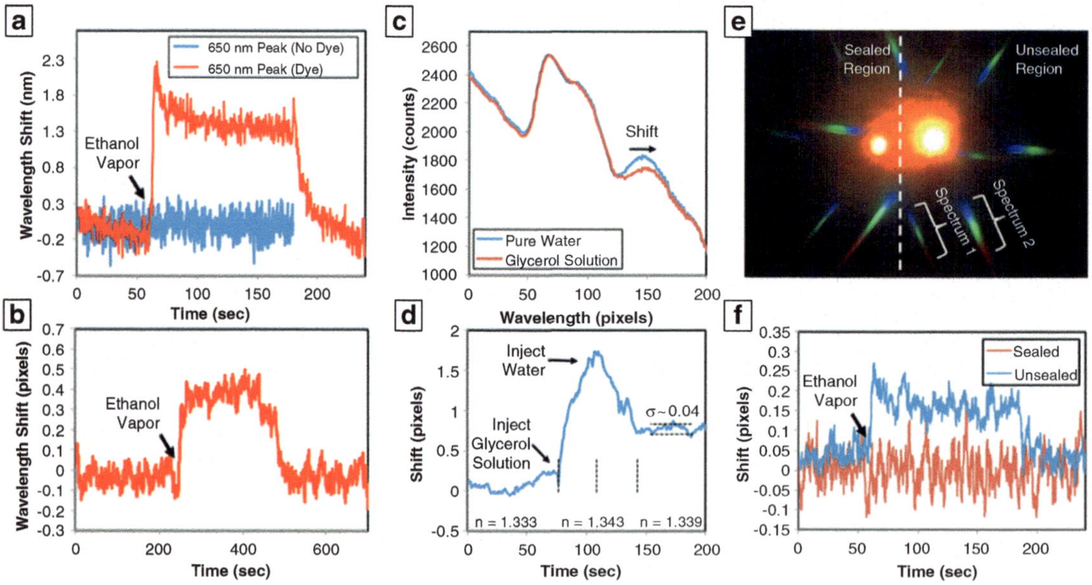

Fig. 3 Sensing with the nanoholes plus imager chip. (**a**) A traditional spectrometer system is used to validate the sensing scheme. Nanohole arrays without the Reichardt's dye did not shift upon exposure to 100 ppm ethanol vapor in nitrogen. (**b**) Sensing data from the nanoholes plus imager system. (**c**) The diffracted transmission peaks are also sensitive to liquid refractive index. (**d**) Real-time sensing experiment with various concentrations of glycerol solution in water. (**e**) Side-by-side sensing spots can be easily created. In this case, half the sensing surface was sealed with tape. (**f**) Upon exposure to ethanol vapor, only the unsealed half responded. Adapted from Ref. 33 with permission from Institute of Physics (IOP) publishing

2 Materials

For Subheadings 2.1 and 2.2, materials should be processed in a Class 100 cleanroom equivalent if possible. It is important to use gloves and personal protective equipment at all times when handling reagents and to follow all waste disposal regulations. Vendors are provided where appropriate.

2.1 Plasmonic Nanohole Array

1. 700 nm Period hexagonal hole silicon nanostamp template (*see* **Note 1**), size 8 × 8.3 × 0.7 mm, groove depth 300 nm (LightSmyth). Groove depth of holes must be greater than deposition thickness (*see* **Note 2**).
2. Chromium etchant: Use caution when handling this acid.
3. Deionized, pure water.
4. Compressed air or dry N_2.
5. Piranha cleaning etchant: 1:1 solution of H_2O_2:H_2SO_4 (*see* **Note 3**). Prepare 100 mL by slowly adding 50 mL H_2O_2 to 50 mL H_2SO_4 in a glass beaker, permitting heat to dissipate. Use caution when handling this acid.
6. Acetone, ≥99%.
7. Isopropyl alcohol, ≥99%.
8. Thermal vacuum evaporator (*see* **Note 4**).
9. Silver evaporation slugs, 99.99% pure.
10. Glass microscope slides, 75 × 25 × 1–1.2 mm.
11. Diamond-tipped glass cutter.
12. Norland 61: Optically clear, UV curable adhesive.
13. UV curing lamp, emitting the long UV range of 320 nm to 400 nm (*see* **Note 5**).

2.2 Solvatochromic Doping

1. Reichardt's dye, solid, at 90% dye: Reconstitute the dye to 10 mM concentration in methanol (*see* **Note 6**).
2. Spin coater capable of 2000+ max rpm with ~25 mm sample mount (roughly ~100 g-force max). The sample was a 25 × 25 mm square glass slide substrate.

2.3 Nanohole Spectrometer Setup

1. Collimated white light source and focusing lens (*see* **Note 7**): All optics were purchased from Thorlabs, Inc.
2. Iris/single-hole aperture (*see* **Note 8**).
3. Vapor-sensing chamber: An airtight cavity with parallel glass windows and ports for analyte input and exhaust (*see* **Note 9**). The sensing chamber was assembled in-house. Use nonreactive, fluoropolymer tubing to bring analytes in and a tube >0.25 m for exhaust.

4. Fiber-optic spectrometer (Ocean Optics): Used only to test and validate the spectral properties of the nanohole array. As is shown, the nanohole array acts as its own spectrometer when placed onto an imager chip.

2.4 Nanohole Imager Chip Setup

1. CMOS imager chip, 1/3″ format, and a 744 × 480 pixel density or greater (*see* **Note 10**): Purchased from The Imaging Source.
2. Microscope immersion oil.
3. Image processing software (*see* **Note 11**): Software was written in-house in LabVIEW.

3 Methods

For Subheadings 3.1 and 3.2, methods should be carried out in a Class 100 cleanroom equivalent or greater if possible. Use gloves and personal protection equipment at all times.

3.1 Nanohole Array Template Stripping Fabrication

1. If reusing the silicon template, submerge it in a beaker of chromium etchant for 30 min to remove previous metallic artifacts (*see* **Note 12**). With tweezers, move the template about the beaker.
2. Remove the template from chromium etchant and immediately rinse with deionized, pure water. Gently blow off remaining water with compressed air until dry.
3. Submerge silicon nanohole template in a beaker of piranha etchant for 30 min. This will prepare the template for silver deposition and template stripping of the patterned nanoholes.
4. Remove the template from piranha etchant and immediately rinse with deionized, pure water. A 10-min soak in water will ensure that any residues are removed. Gently blow off remaining water with compressed air until dry.
5. If any artifacts, metallic or organic, remain on your nanohole template, and repeat **steps 1–4** (*see* **Note 13**).
6. Load the thermal deposition chamber with silver evaporation slugs and secure the template within the chamber. Do not use a sputter coater since the film deposition needs to be directional. Deposit 50 nm of silver at a rate of 10 nm per minute (*see* **Note 14**). Thicker films may provide better optical properties, but the hole may become plugged. The film thickness needs to be less than the depth of the template holes to create a discontinuous deposition (*see* Fig. 1a).
7. Score a 3 × 1 in. microscope slide with the diamond-tipped glass cutter and cleave into three 1 × 1 in. glass slides. The two

additional slides may be stowed or used in parallel for batch fabrication.

8. Clean the microscope slide. Rinse with acetone and then isopropyl alcohol followed by deionized, pure water. Dry in a stream of air.
9. Place a small aliquot of Norland 61 UV curable adhesive, on the order of 1 μL, onto the silver-coated nanohole template (*see* **Note 15**). Place the 1 × 1 in. glass slide centered on top of the template, compressing the aliquot of adhesive. Do not reposition the slide or adjust the adhesive after initial contact.
10. Carefully load the template, adhesive, and glass plate onto the UV curing stage and cure the adhesive through the glass slide for roughly 10 min.
11. Turn the glass slide over so the nanohole template is on top. Strip the deposited silver nanohole array away from the template with controlled leverage with a razor blade from the side of the template. Apply a force at the interface between the template and the glass. The template will separate from the glass, pulling the silver nanoholes with it. This happens because the adhesion of the adhesive to the silver is greater than the adhesion of the silver to the silicon template (*see* **Note 16** and Fig. 1b).

3.2 Doping the Nanohole Array with Reichardt's Solvatochromic Dye

1. If the nanoholes are to be used for vapor sensing of volatile organic compounds in air, following the template stripping procedure, affix the nanohole array (and glass slide) upright and centered on the spin coater.
2. Pipette 75.0 μL 10 mM Reichardt's dye in methanol directly onto the silver nanohole array and let rest for 30 s. Spin at 100 × *g* for 30 s. The radius of the sample mount was ~25 mm to accommodate a centered 25 × 25 mm square glass slide substrate.

3.3 Plasmonic Nanohole Array Sensing with Traditional Spectrometer

This approach resembles a more traditional plasmonic sensing setup with coupled transmission from the nanohole array entering a fiber-optic spectrometer. This setup is useful to verify spectral shifts and to attain higher spectral resolution. Various microfluidic chambers can be used.

1. Bring the silver nanohole chip into a sensing chamber by affixing the glass slide to the floor of the chamber with the nanohole sample inside the cavity.
2. Seal the perimeter of the glass slide outside the chamber so the analyte input and exhaust are the only air or liquid transferred into the chamber.

3. Use inert fluoropolymer tubing to extend the analyte input to the source and connect a small length of tubing from the chamber's exhaust into a waste zone. Follow all chemical hazard documentation of analyte to determine the suitable degree of controlled waste management.
4. Direct collimated white light with a condenser lens through the chamber window onto the nanohole sample. The numerical aperture of the condenser should remain low enough that the illumination remains roughly collimated. The illumination should appear as a single point of light on the surface of the nanohole sample and will transmit through the glass on the reverse side.
5. Align the fiber optic to couple the transmitted spectrum of the nanohole array sample into the spectrometer (*see* **Note 17**). A coupling close to the test chamber and encased in a light-blocking sheath aids in reducing stray light input. Peaks in this spectrum read from the spectrometer correlate to resonances of the plasmonic substrate (*see* **Note 18**).
6. Introduce the vapor or liquid analyte into the sensing chamber's input tube (*see* **Note 19**).
7. Track relative shifts in wavelength of the transmission peaks from spectrometer (*see* **Note 20**).

3.4 Direct Spectral Imaging of Plasmonic Nanohole Array Sensing on CMOS Setup

This approach measures the first-order diffraction directly on a CMOS imager chip, removing the necessity of an external spectrometer. This setup can also be expanded to multiplex sensing, not possible with a single fiber-optic spectrometer (*see* **Note 21**).

1. Follow **steps 1** and **2** of Subheading 3.3 to align the excitation light onto the nanohole sample.
2. Apply a drop of microscope immersion oil on the CMOS imager, enough to cover the pixel array. Note that the CMOS camera does not have a lens. You will need direct access to the imager surface (covered with a protective glass window; *see* Fig. 2).
3. Affix the CMOS imager under the vapor sensing chamber so the oil contacts the glass slide substrate underneath the nanohole array.
4. Power the broadband LED and image the diffraction spectrum on the CMOS imager. Tune the distance between the glass slide and the imager so that the first-order diffraction pattern is contained completely within the imaging boundary of the CMOS imager (*see* Fig. 2c).
5. Measure the first-order hexagonal diffraction spectrum and calibrate for sensing (*see* **Note 22**).

6. Introduce the analyte to the sensing chamber (*see* **Note 19**).
7. Track relative shifts across pixels of the first-order spectrum peaks of the CMOS imager in real time (*see* **Notes 23** and **24**).

4 Notes

1. Nanotemplates were purchased from LightSmyth Technologies, Inc. where a wide variety of nanopatterns are available for selection. A template with a hexagonal lattice of holes at a 700 nm period was determined through experimentation and simulation to provide the effect of a diffraction grating as well as transmission via plasmon-mediated extraordinary optical transmission. Incident light upon the grating emerged beyond the critical angle at six points as a result of the hexagonal nano-features. If not for the immersion oil, this light is trapped by total internal reflection within the glass substrate.
2. Scanning electron microscope images showed the formation of metal "plugs" in the holes of the resultant template-stripped substrate when deposition thickness neared or exceeded the groove depth of nano-features. A margin of *ca.* 100 nm should be kept to ensure the retention of nanohole features. Our nanohole template had features that were 300 nm deep.
3. Piranha etchant is a powerful solution to remove organic matter from surfaces of the substrate, ideal for cleaning nano-featured silicon. Extreme caution should be taken when preparing and interacting with the solution. During preparation, the solution is highly exothermic and should be made by adding H_2O_2 to H_2SO_4 slowly as to minimize volatile release of heat and corrosive fumes. We find it best to prepare the solution fresh.
4. We selected the *VapourStation* benchtop thermal evaporator (Oxford Vacuum Science) for its high vacuum: 10^{-5}–10^{-9} mbar, integrated lid mounting stage, shutter, and quartz crystal deposition monitor head.
5. Norland 61 is a UV curable adhesive with maximum absorption within the range of 320–380 nm with peak sensitivity around 365 nm. Full cure requires 3 J cm^{-2}. A UV source that fulfills these parameters will ensure optimal curing. The UV lamp used in the experiments was a 365 nm Thorlabs mounted LED with a 1150 mW output power.
6. Reichardt's dye is a prominent solvatochromic dye, notable for its diverse range of reactive properties, and was therefore selected to enhance the vapor detection capabilities of the plasmonic nanoholes. Reconstitution of the dye as a 10 mM solution was tested in different four solvents: acetone,

isopropanol, methanol, and ethanol. Methanol was found to be the most effective and consistent carrier for application onto the nanohole array surface.

7. The white light source used in the setup was a broadband white LED (Thorlabs) that was roughly collimated using a condenser lens. An aperture limits the beam waist of collimated light. The focusing lens, 75 cm focal length (Thorlabs), is used to further reduce the beam waist, focus the light onto the nanohole surface, and retain near-normal incidence.
8. We have used a threaded iris (Thorlabs) to achieve an aperture of approximately 0.5 mm. Alternatively, a needle hole through blackout aluminum also proved an effective aperture and was later used with an array of punctures as a multi-hole aperture for multiplexed sensing. An important note is that the focused light should be as small as possible to maintain good spectral resolution.
9. The vapor-sensing chamber is customizable and should be tailored to fit the system and intended use. We constructed a collimated white light source with Thorlabs' 30 mm cage system. To conveniently align the sensing chamber in the same cage system, we machine milled an approximate 10 × 10 mm cavity through the center of a blank 30 mm cage plate and tapped two ports on the sides to affix vacuum tubing with screw-in barbs. The milled cavity was sealed with a permanent glass microscope slide window and O-ring gasket on top and a quick-release microscope slide and O-ring gasket on bottom. A quick release was selected on bottom for the ease of placing the plasmonic nanohole array.
10. A single-board-color CMOS imager chip (The Imaging Source) was used to produce first-order diffraction images; however, a black and white CMOS imager chip was used in the sensing setup. The 744 × 480 pixel in a 1/3 in. format corresponds to 6 μm pixels and is the fundamental limitation of spectral resolution. Improvements to spectral resolution, signal to noise, and overall device performance include a higher density CMOS pixel array for increased spectral resolution, an imager with more than 8 bits of digitization, and an imager with higher dynamic range to avoid saturation.
11. Focused light upon the hexagonal nanoholes produces six first-order diffraction bands that are imaged on the camera chip (*see* Fig. 2c). We used a homemade LabVIEW interface to the CMOS imagers to extract a cross-sectional line profile of a band's intensity data from the imager in real time. The cross-sectional intensity data showed three peaks which were used for tracking during vapor sensing.

12. Of the various nano-fabrication techniques available, template stripping offers many significant advantages: notably ultra-smooth surfaces, long plasmon propagation lengths, and ability to pattern arrangements of bumps, holes, and grooves over large areas with reusable templates. Reusing nanohole templates in consecutive depositions requires removal of residual silver from previous depositions. Inadequate removal of previously deposited metal can result in irregular optical surfaces and a decreased performance. With continuous reuse, it is possible to scratch or degrade the surface, after which the template should be retired. But with care, it is possible to successfully use the same template dozens of times.
13. Highly contaminated samples may require further rounds of more thorough cleaning.
14. A continuous deposition ensures high-fidelity nano-features. If possible, block the template from stray deposition with a shutter as the chamber decompresses, heats, and ramps to the rate of 10 nm per minute.
15. An efficient method of dispensing the small aliquot of Norland 61 is to dip the tip of a needle or syringe into a drop of the adhesive and gently touch the silver-deposited surface of the nanohole template. Err on the small side in regard to volume, since too much adhesive can easily spread around the edges of the template (i.e., not on the silver) and form a strong bond between the glass substrate and the template, risking damage upon stripping.
16. We found it best to use a razor blade to template strip. Before applying force, place the blade on the edge of the adhered template at approximately 30° with respect to the glass. Carefully apply progressively incrementing force. If properly cleaned, the silver will easily cleave from the silicon template. When the template does clear, use caution as to not slide the edge of the blade into the silver nanohole array. If the template is too firmly adhered or incorrect force is applied, the silicon may crack, resulting in a broken template.
17. Light passes through the nanohole arrays via plasmon-mediated extraordinary optical transmission. Shifts in the transmission peaks are used for refractive index sensing [13].
18. The free-space wavelengths of plasmon-mediated transmission maxima λ_{max} can be estimated via a grating coupling mechanism for a hexagonal lattice and illumination at normal incidence. With a hexagonal period of 700 nm and the dielectric constant of the silver fit to a Drude model, substrate-side resonances are predicted at approximately 629, 606, and 545 nm. The grating coupling equation does not account for the specific hole shape, film thickness, surface roughness,

rounded hole edges, or other more complicated processes that lead to the enhanced transmission effect. See refs. 33 and 34 for more information.

19. For the vapor analyte sensing, we used a 99.99% pure nitrogen tank source connected to a mass flow controller (Cole Parmer), a set of three-way valves (Cole Parmer), and a homemade mixing chamber that would receive liquid ethanol from a syringe pump (Cole Parmer). This allowed us to vary the vapor concentration of different solvents in nitrogen. Characteristic spectral shifts in the transmission spectrum can be attributed to the presence of different analytes with the magnitude of the shift correlating to concentration.
20. Raw spectrometer data showed a series of three distinct transmission peaks which shift in wavelength with response to analyte introduction (Fig. 1e). The peaks that correspond to a resonance on the sensing surface (i.e., air or liquid) will shift with changes in refractive index. Other peaks will not shift, unless the nanohole substrate (i.e., the adhesive) is specifically fabricated to shift upon exposure to the analyte [33, 34]. The peaks were typically fit with a local third-order polynomial and tracked over time. The sensor surface is typically flushed with pure nitrogen (for vapor sensing) or pure water (for liquid index sensing) for several minutes to establish a baseline. The analytes are then introduced at various concentrations for several minutes. The surface is then flushed again, returning to the baseline or to another level that corresponds to any permanent adhesion of analytes to the surface.
21. Modifying the single-hole aperture to an array of apertures will illuminate several locations on the nanohole sample at once for multiplexed sensing (Fig. 3e) [24].
22. The undiffracted zero-order light can oversaturate the imager, obscuring the first-order diffraction spectrum bands from observation. The aperture size was considerably reduced from the spectrometer setup to minimize saturation. The exposure time of the imager was also modified so that six distinct diffraction bands can be viewed clearly. An imager with a greater dynamic range than the one used in the experiment would also help maintain a clearer signal.
23. The intensity data of the six first-order diffraction bands can be viewed directly on the CMOS imager. The setup used external image processing to extract pixel intensity data across the imager and track sensing effects in real time. Data from the bands was gathered by taking a line profile from the cross section of a single band. Each diffraction pattern band shows an intensity profile with the characteristic three-peak pattern observed in the transmission spectra. A third-order polynomial

fitting function was applied to track the CMOS pixel location of the relevant peaks.

24. Sensitivity to liquid immersion CMOS imager and nanohole sensing platform shows general applicability of the direct spectral imaging approach. Nanohole chips were fabricated and immersed in 20 μL drop of water ($n = 1.333$) placed directly on the nanohole surface in the sample chamber. Extracting the spectrum from the CMOS pixels produced a series of peaks, one of which shifted when mixed with a 20 μL drop of water and glycerol, increasing the refractive index of the solution ($n = 1.343$). Tracking this peak in real time shows response of the peak to these changes in refractive index (Fig. 3c, d). Upon a final addition of another 10 μL drop of water, the solution had a slightly decreased refractive index ($n = 1.339$) and finally stabilized (Fig. 3d).

Acknowledgments

The authors thank Phil Minell and James Myrick (*US Enginewing*) for collaborative discussion and support. Parts of this research were also supported by the Minnesota Space Grant Consortium (MnSGC), part of the NASA-funded National Space Grant College and Fellowship Program.

References

1. Sia SK, Kricka LJ (2008) Microfluidics and point-of-care testing. Lab Chip 8:1982–1983
2. Chin CD, Laksanasopin T, Cheung YK, Steinmiller D, Linder V, Parsa H, Wang J, Moore H, Rouse R, Umviligihozo G et al (2011) Microfluidics-based diagnostics of infectious diseases in the developing world. Nat Med 17:1015–1019
3. Homola J, Yee SS, Gauglitz G (1999) Surface plasmon resonance sensors: review. Sensor Actuat B-Chem 54:3
4. Cooper MA (2002) Optical biosensors in drug discovery. Nat Rev Drug Discov 1:515
5. Whittle CL, Fakharzadeh S, Eades J, Preti G (2007) Human breath odors and their use in diagnosis. Ann N Y Acad Sci 1098:252–266
6. Biggs KB, Camden JP, Anker JN, Duyne RPV (2009) Surface-enhanced Raman spectroscopy of benzenethiol adsorbed from the gas phase onto silver film over nanosphere surfaces: determination of the sticking probability and detection limit time. J Phys Chem A 113:4581–4586
7. Stuart DA, Biggs KB, Duyne RPV (2006) Surface-enhanced Raman spectroscopy of half-mustard agent. Analyst 131:568–572
8. Kahn N, Lavie O, Paz M, Segev Y, Haick H (2015) Dynamic nanoparticle-based flexible sensors: diagnosis of ovarian carcinoma from exhaled breath. Nano Lett 15:7023–7028
9. Hodgkinson J, Tatam RP (2013) Optical gas sensing: a review. Meas Sci Technol 24:012004
10. Long F, Zhu A, Shi H (2013) Recent advances in optical biosensors for environmental monitoring and early warning. Sensors 13:13928–13,948
11. Anker JN, Hall WP, Lyandres O, Shah NC, Zhao J, Van Duyne RP (2008) Biosensing with plasmonic nanosensors. Nat Mater 7:442–453
12. Stewart ME, Anderton CR, Thompson LB, Maria J, Gray SK, Rogers JA, Nuzzo RG (2008) Nanostructured plasmonic sensors. Chem Rev 108:494–521
13. Gordon R, Sinton D, Kavanagh KL, Brolo AG (2008) A new generation of sensors based on

extraordinary optical transmission. Acc Chem Res 41:1049–1057
14. Barnes WL, Dereux A, Ebbesen TW (2003) Surface plasmon subwavelength optics. Nature 424:824
15. Bingham JM, Anker JN, Kreno LE, Van Duyne RP (2010) Gas sensing with high-resolution localized surface plasmon resonance spectroscopy. J Am Chem Soc 132:17358–17359
16. Ebbesen TW, Lezec HJ, Ghaemi HF, Thio T, Wolff PA (1998) Extraordinary optical transmission through sub-wavelength hole arrays. Nature 391:667–669
17. De Leebeeck A, Kumar LKS, de Lange V, Sinton D, Gordon R, Brolo AG et al (2007) On-chip surface-based detection with nanohole arrays. Anal Chem 79:4094–4100
18. Tetz KA, Pang L, Fainman Y (2006) High-resolution surface plasmon resonance sensor based on linewidth-optimized nanohole array transmittance. Opt Lett 31:1528
19. Brolo AG, Gordon R, Leathem B, Kavanaghs KL (2004) Surface plasmon sensor based on the enhanced light transmission through arrays of nanoholes in gold films. Langmuir 20:4813–4815
20. Im H, Lesuffleur A, Lindquist NC, Oh SH (2009) Plasmonic nanoholes in a multichannel microarray format for parallel kinetic assays and differential sensing. Anal Chem 81:2854
21. Lindquist NC, Lesuffleur A, Im H, Oh SH (2009) Sub-micron resolution surface plasmon resonance imaging enabled by nanohole arrays with surrounding Bragg mirrors for enhanced sensitivity and isolation. Lab Chip 9:382–387
22. Lee SH, Lindquist NC, Wittenberg NJ, Jordan LR, Oh S (2012) Real-time full-spectral imaging and affinity measurements from 50 microfluidic channels using nanohole surface plasmon resonance. Lab Chip 12:3882–3890
23. Eftekhari F, Escobedo C, Ferreira J, Duan X, Girotto EM, Brolo AG, Gordon R, Sinton D (2009) Nanoholes as nanochannels: flow-through plasmonic sensing. Anal Chem 81:4308–4311
24. Escobedo C, Brolo AG, Gordon R, Sinton D (2010) Flow-through vs. flow-over: analysis of transport and binding in nanohole array plasmonic biosensors. Anal Chem 82:10015–10020
25. Liedberg B, Nylander C, Lundstrom I (1983) Surface plasmon resonance for gas detection and biosensing. Sensor Actuator 4:299
26. Nylander C, Liedberg B, Lind T (1983) Gas detection by means of surface plasmon resonance. Sensor Actuator 3:79–88
27. Vukusic P, Bryan-Brown G, Sambles J (1992) Surface plasmon resonance on gratings as a novel means for gas sensing. Sensors Actuat B-Chem 8:155–160
28. Miwa S, Arakawa T (1996) Selective gas detection by means of surface plasmon resonance sensors. Thin Solid Films 281:466–468
29. Notcovich AG, Zhuk V, Lipson S (2000) Surface plasmon resonance phase imaging. Appl Phys Lett 76:1665–1667
30. Wright JB, Cicotte KN, Subramania G, Dirk SM, Brener I (2012) Chemoselective gas sensors based on plasmonic nanohole arrays. Opt Mater Express 2:1655–1662
31. Chen Y, Lu C (2009) Surface modification on silver nanoparticles for enhancing vapor selectivity of localized surface plasmon resonance sensors. Sensors Actuat B-Chem 135:492–498
32. Ma W, Yang H, Wang W, Gao P, Yao J (2011) Ethanol vapor sensing properties of triangular silver nanostructures based on localized surface plasmon resonance. Sensors 11:8643–8653
33. Seiler ST, Rich IS, Lindquist NC (2016) Direct spectral imaging of plasmonic nanohole arrays for real-time sensing. Nanotechnology 27:184001
34. Lindquist NC, Turner MA, Heppner BP (2014) Template fabricated plasmonic nanoholes on analyte-sensitive substrates for real-time vapor sensing. RSC Adv 4:15115–15,121
35. Im H, Lee SH, Wittenberg NJ, Johnson TW, Lindquist NC, Nagpal P, Norris DJ, Oh SH (2011) Template-stripped smooth Ag nanohole arrays with silica shells for surface plasmon resonance biosensing. ACS Nano 5:6244
36. Cetin AE, Coskun AF, Galarreta BC, Huang M, Herman D, Ozcan A, Altug H (2014) Handheld high-throughput plasmonic biosensor using computational on-chip imaging. Light-Sci Appl 3:e122
37. Reichardt C (1994) Solvatochromic dyes as solvent polarity indicators. Chem Rev 94:2319–2358
38. Blum P, Mohr GJ, Matern K, Reichert J, Spichiger-Keller UE (2001) Optical alcohol sensor using lipophilic Reichardt's dyes in polymer membranes. Anal Chim Acta 432:269–275
39. Sadaoka Y, Sakai Y, Murata Y (1992) Optical humidity and ammonia gas sensors using Reichardt's dye-polymer composites. Talanta 39:1675–1679

Chapter 9

Inkjet-Printed Colorimetric Paper-Based Gas Sensor Arrays for the Discrimination of Volatile Primary Amines with Amine-Responsive Dye-Encapsulating Polymer Nanoparticles

Hiroyuki Shibata, Yuma Ikeda, and Daniel Citterio

Abstract

Arrays of gas sensors are of high interest for "electronic nose" applications. Here, we describe the fabrication of a colorimetric single-use gas sensor array made of paper allowing the discrimination of volatile primary amines based on their polarity. For this purpose, polymeric nanoparticles with different polarities containing an amine-sensitive chromogenic dye are deposited onto paper substrates by means of inkjet printing. Data processing is conducted by red-green-blue (RGB) color extraction, followed by principal component analysis (PCA) or agglomerative hierarchical clustering (AHC) analysis. The application to the discrimination of two cheese samples is demonstrated.

Key words Colorimetric gas sensor array, Paper-based analytical devices, Inkjet-printing, Volatile primary amines, Dye-encapsulating polymer nanoparticles, Principal component analysis, Agglomerative hierarchical clustering

1 Introduction

Colorimetric gas sensor arrays have drawn remarkable attention in the context of electronic noses, providing a facile, efficient, and sensitive approach for the rapid detection and identification of a wide variety of volatile compounds in various fields of application [1]. In the food and beverage industry for example, preferred analytical techniques for quality control in the raw material supply chain are approaches allowing for easy and rapid comparison with reference samples, rather than those relying on significantly more expensive and work-intensive large analytical instrumentation (*e.g.*, gas and liquid chromatography, mass spectrometry). Suslick and co-workers originally introduced a very simple and low-cost type of sensor array ideally suited for electronic nose applications by depositing multiple chemically responsive indicators (*e.g.*, pH

Jessica E. Fitzgerald and Hicham Fenniri (eds.), *Biomimetic Sensing: Methods and Protocols*, Methods in Molecular Biology, vol. 2027, https://doi.org/10.1007/978-1-4939-9616-2_9,

indicators, solvatochromic dyes, redox indicators, metalloporphyrin derivatives) as individual spots on a hydrophobic substrate [2–4]. Exposure of the sensor array to gaseous analytes produces analyte-specific color change patterns. Several research groups have successfully introduced variations of this principle in the form of colorimetric sensor arrays based on various optically responsive dyes [5–9].

In this protocol, we describe the fabrication procedure for paper-based colorimetric sensor arrays by means of inkjet printing and their application for discrimination of volatile primary alkylamines with different alkyl chain lengths [6]. In addition, the discrimination of two types of cheeses by headspace gas analysis using the paper-based sensor array is demonstrated as practical application example.

Paper has played an important role in chemical assays for many years, due to its multiple advantages that include (1) low cost, (2) disposability by incineration, and (3) portability [10–14]. To achieve reproducible, flexible, and affordable sensor fabrication [15], inkjet-printing technology has been applied for the deposition of two types of polymeric nanoparticles of different polarity, containing a chemically functional dye with selectivity for amines in general, onto standard copy paper as the sensor substrate. The capability to distinguish one closely related amine from others is achieved by a polarity-based approach. The resulting paper-based sensor arrays are not influenced by environmental humidity and do not respond to other common volatile organic compounds [6]. For processing of red-green-blue (RGB) color space colorimetric raw data collected by a scanner, multivariate statistical analysis such as principal component analysis (PCA) and agglomerative hierarchical clustering (AHC) have been applied.

2 Materials

2.1 Synthesis of Amine-Responsive Dye

All solvents and reagents for organic synthesis are available from commercial suppliers and can be used without further purification. Strictly follow local waste disposal regulations when disposing waste materials:

1. Aniline (starting material).
2. 1-Bromohexane.
3. *N*-Ethyldiisopropylamine (DIPEA).
4. Dimethylformamide.
5. Chloroform.
6. Anhydrous sodium sulfate (Na_2SO_4).
7. 4-Iodoaniline.

8. Hydrochloric acid.
9. Sodium nitrite ($NaNO_2$).
10. Ethanol.
11. Sodium bicarbonate ($NaHCO_3$).
12. Diethyl ether.
13. 1.6 M *n*-Butyl lithium solution in hexane.
14. Ethyl trifluoroacetate.
15. Dichloromethane.
16. Argon gas protective atmosphere.
17. 100 mL Three-necked recovery flask.
18. 30 mL Two-necked recovery flask.
19. 50 mL Two-necked recovery flask.
20. Flash chromatography column: Silica gel, eluent composition: hexane:ethyl acetate = 95:5–50:50).
21. Second chromatography column: Silica gel, eluent composition: hexane:chloroform = 75:25.
22. Third chromatography column: Silica gel, eluent composition: hexane:chloroform = 66:33–50:50.
23. Hotplate.
24. JEOL-ECA-500 MHz NMR spectrometer.
25. Xevo G2 QTof series mass spectrometer.

2.2 Preparation of Polymer Particles

All reagents are available from commercial suppliers. Monomers are purified before use by passing them through a layer of aluminum oxide, while all other reagents can be used as received:

1. Benzyl methacrylate (BzMA, monomer).
2. Diethylene glycol methyl ether methacrylate (DEGMMA, monomer).
3. Methyl methacrylate (MMA, monomer).
4. Sodium dodecyl sulfate (SDS, surfactant).
5. Sodium bicarbonate ($NaHCO_3$).
6. Potassium persulfate (KPS, polymerization initiator).
7. Nitrogen gas.
8. 50 mL Four-necked recovery flask.
9. Hotplate with mechanical stirrer.
10. Syringe.
11. Dialysis tubing.

2.3 Dye Encapsulation into Polymer Nanoparticles

1. 200 mL Recovery flask.
2. Tetrahydrofuran (THF).
3. Pipettes.

2.4 Fabrication of Inkjet-Printed Paper-Based Sensor Array

All materials are available from commercial suppliers:

1. Ethylene glycol (printing ink viscosity modifier).
2. Standard white copy paper (size A4, sensor array substrate).
3. Flasks (to prepare ink solutions).
4. Syringe.
5. 0.22 μm Syringe filter.
6. 10 pL Nominal droplet volume inkjet printing cartridge of a Fujifilm Dimatix DMP-2831 material printer (six total).
7. Inkjet printer.
8. Desiccator with vacuum.

2.5 Volatile Primary Alkylamines for Sensor Array Response Characterization

All amines are available from commercial suppliers and can be used without further purification:

1. *n*-Methylamine.
2. *n*-Ethylamine.
3. *n*-Propylamine.
4. *n*-Butylamine.
5. *n*-Pentylamine.
6. *n*-Hexylamine.
7. *n*-Heptylamine.
8. Argon gas.
9. Bubbler with a sealed container.
10. Heat gun.
11. Gas sampling bags.
12. Flow-through measurement chamber (*see* Fig. 3).
13. Pump.
14. Canon CanoScan 5600F color scanner.
15. ImageJ software.

3 Methods

3.1 Synthesis of the Amine-Responsive Dye

As the chromogenic sensing element, a derivative of the amine-responsive azo dye ETH^{T} 4001 (chromoreactant) has been selected [16–20]. The synthesis scheme is shown in Fig. 1.

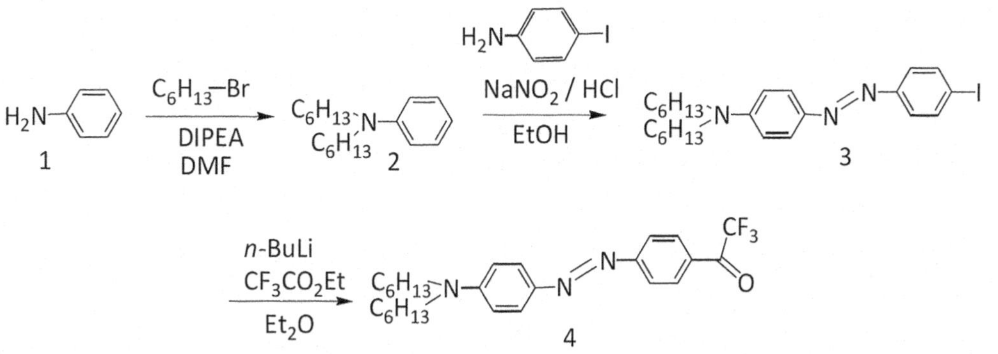

Fig. 1 Synthesis scheme for the amine-responsive azo dye. Reprinted with permission from Soga T, Jimbo Y, Suzuki K, Citterio D, Anal. Chem., 2013, 85, 8973-8978 Copyright (2013) American Chemical Society (**6**)

3.1.1 General Procedure for Synthesis

1. Carry out all moisture-sensitive reactions under an argon gas protective atmosphere.
2. The composition of mixed solvents is indicated as the volume ratio (v:v).
3. Record NMR spectra on a JEOL-ECA-500 MHz NMR spectrometer at room temperature at 500 MHz (^{1}H-NMR), 125 MHz (^{13}C-NMR), and 471 MHz (^{19}F-NMR), respectively.
4. Chemical shifts are indicated relative to an internal standard of tetramethylsilane (δ = 0.0 ppm) for ^{1}H, trifluoroacetic acid (δ = −75.0 ppm) for ^{19}F, or the solvent residual peak ($CDCl_3$: δ = 77.1 ppm) for ^{13}C. Coupling constants are given in Hz.
5. Record electrospray ionization mass spectra (ESI-MS) on a Xevo G2 QTof series mass spectrometer.

*3.1.2 Synthesis of N,N-dihexylaniline (See Fig. 1, Compound (**2**))*

1. In a 100 mL three-necked recovery flask, dissolve aniline (5.25 g, 56.4 mmol, 1.0 eq.), 1-bromohexane (25.0 g, 0.151 mol, 2.7 eq.), and *N*-ethyldiisopropylamine (22.2 g, 0.171 mol, 3.0 eq.) into dimethylformamide (30 mL) and stir at 110 °C for 1 h.
2. After cooling to room temperature, pour the reaction mixture into distilled water, and extract three times with chloroform.
3. Wash the combined organic phases two times with water, dry over Na_2SO_4, and remove the solvent by evaporation.
4. Purify the resulting dark brown oil by flash chromatography (silica gel, eluent composition: hexane:ethyl acetate = 95:5–50:50):

   Yield: 13.8 g (94%, yellow oil) (*see* **Note 1**). ^{1}H-NMR (500 MHz, $CDCl_3$, TMS): δ (ppm) = 0.89 (m, 6H), 1.31 (m, 12H), 1.56 (m, 4H), 3.24 (t, 4H, J = 7.7 Hz), 6.62 (m, 3H), 7.18 (m, 2H).

3.1.3 Synthesis of 4-(N,N-dihexylamino)-4′-iodoazobenzene (See Fig. 1, Compound (3))

1. In a 30 mL two-necked recovery flask, suspend 4-iodoaniline (830 mg, 3.79 mmol, 1.0 eq.) in 6 M hydrochloric acid (3.0 mL) and cool to 5 °C in an ice bath.
2. Add $NaNO_2$ (255 mg, 3.69 mmol, 0.97 eq.) dissolved in distilled water (2 mL) into the suspension.
3. After filtration, add the resulting solution in a dropwise manner to *N*,*N*-dihexylaniline (**2**) (928 mg, 3.55 mmol, 0.94 eq.) in ethanol (20 mL) containing concentrated hydrochloric acid (0.5 mL) and stir for 3.5 h in an ice bath.
4. Remove the solvent by evaporation and take up the resulting residue in chloroform.
5. Wash the organic phase once with aqueous $NaHCO_3$ solution and three times with water, dry over Na_2SO_4, and remove the solvent by evaporation.
6. Purify the resulting purple oil by column chromatography (silica gel, eluent composition: hexane:chloroform = 75:25).

   Yield: 615 mg (33%, red oil). ^{1}H-NMR (500 MHz, $CDCl_3$, TMS) δ (ppm) = 0.91 (m, 6H), 1.34 (m, 12H), 1.63 (m, 4H), 3.35 (t, 4H, $J = 7.7$ Hz), 6.66 (d, 2H, $J = 9.5$ Hz), 7.56 (d, 2H, $J = 8.9$ Hz), 7.79 (d, 2H, $J = 8.9$ Hz), 7.82 (d, 2H, $J = 9.2$ Hz).

3.1.4 Synthesis of 4-dihexylamino-4′-(trifluoroacetyl)azobenzene (See Fig. 1, Compound (4))

1. In a 50 mL two-necked recovery flask, dissolve 4-(*N*,*N*-dihexylamino)-4′-iodoazobenzene (**3**) (208 mg, 0.422 mmol, 1.0 eq.) in diethyl ether (11.0 mL) and stir at −78 °C under an atmosphere of argon gas.
2. Dropwise add 1.6 M butyl lithium hexane solution (0.4 mL, 0.64 mmol, 1.5 eq.) into the reaction solution and stir at −78 °C for 10 min.
3. Slowly add ethyl trifluoroacetate (0.8 mL, 6.7 mmol, 16.0 eq.) into the reaction mixture and continue stirring at −78 °C for 2.5 h.
4. Add aqueous $NaHCO_3$ solution into the mixture.
5. Extract three times with dichloromethane, dry over Na_2SO_4, and remove the solvent by evaporation.
6. Purify the resulting crude product by column chromatography (silica gel, eluent composition: hexane: chloroform = 66:33–50:50).

   Yield: 62.9 mg (32%, dark red solid). ^{1}H-NMR (500 MHz, $CDCl_3$, TMS) δ (ppm) = 0.91 (t, 6H, $J = 7.0$ Hz), 1.35 (br, 12H), 1.65 (br, 4H), 3.38 (t, 4H, $J = 7.8$ Hz), 6.70 (d, 2H, $J = 9.5$ Hz), 7.90 (m, 4H), 8.17 (d, 2H, $J = 8.1$ Hz). ^{13}C-NMR (125 MHz, $CDCl_3$, TMS) δ (ppm) = 14.1, 22.8, 26.8, 27.4, 31.8, 51.4, 111.3,

115.8, 118.1, 122.6, 126.4, 129.2, 131.4, 143.5, 151.7, 157.6, 179.8. ^{19}F-NMR (471 MHz, $CDCl_3$, trifluoroacetic acid as internal standard) δ (ppm) = −72.3 (s). ESI-MS: calculated for $C_{26}H_{34}F_3N_3O$: m/z 461.27, found 462.27 $[M+H]^+$ and 480.28 $[M+H_2O+H]^+$.

3.2 Synthesis of Polymer Particles

1. In a 50 mL four-necked recovery flask, dissolve SDS (0.184 g) and $NaHCO_3$ (7.0 mg) in water (27 mL).
2. Add 4.0 g of monomer (BzMA) into the aqueous mixture and stir at room temperature for 10 min to obtain a homogenous solution.
3. Heat to 80 °C, stirring at 300 rpm with a mechanical stirrer for 30 min, while purging with nitrogen gas.
4. Inject the polymerization initiator solution [KPS (21 mg) in water (5 mL)] into the reaction vessel and continue stirring at 80 °C for 4 h.
5. After cooling to room temperature, transfer the resulting suspension into a dialysis tube and dialyze against water for 3–5 days to remove excess surfactant, affording poly(benzyl methacrylate) (pBzMA) nanoparticles.
6. By using the same procedure as described in **steps 1–5** above for pBzMA, poly(diethylene glycol methyl ether methacrylate-co-methyl methacrylate) [p(DEGMMA-co-MMA)] nanoparticles are prepared by using a mixture of DEGMMA (2.4 g) and MMA (1.6 g) as the monomers instead of BzMA in **step 2**.

3.3 Dye Encapsulation into Polymer Nanoparticles

Encapsulation of the amine-responsive dye 4-dihexylamino-4′-(trifluoroacetyl)-azobenzene (Fig. 1, Compound (**4**)) into the polymeric particles is conducted similarly to a previously published procedure [21]. The procedure is identical for the encapsulation of the dye into either the pBzMA or the p(DEGMMA-co-MMA) nanoparticles.

1. Prepare a 200 mL recovery flask containing a certain amount of the respective polymer particle emulsion (corresponding to 200 mg polymer solid).
2. Add a mixture of distilled water (50 mL) and THF (10 mL), followed by stirring for a few minutes.
3. Prepare a solution of the amine-responsive dye (5.0 mg) in THF (10 mL).
4. Slowly add the dye solution in a dropwise manner to the stirred polymer emulsion.
5. After complete addition of the dye solution, evaporate the THF solvent and concentrate the emulsion under reduced pressure.

6. Dilute the residue with distilled water up to 5 mL overall volume, resulting in a nanoparticle emulsion with a polymer solid content of 40 mg/mL.

3.4 Fabrication of Inkjet-Printed Paper-Based Sensor Arrays

3.4.1 Preparation of Printing Inks

1. Combine a total of 250 μL of the two types of dye-encapsulating polymer nanoparticle emulsions [pBzMA or p(DEGMMA-co-MMA)] in six different mixing ratios: pBzMA:p(DEGMMA-co-MMA) (v:v) = 100:0, 80:20, 60:40, 40:60, 20:80, 0:100.
2. Add 650 μL of water and 100 μL of ethylene glycol, resulting in a polymer solid content of 10 mg/mL in water with 10% (v/v) of ethylene glycol.

3.4.2 Inkjet-Printing of Nanoparticle Inks

1. Using a syringe, inject the printing ink through a 0.22 μm syringe filter into a 10 pL nominal droplet volume inkjet-printing cartridge of a Fujifilm Dimatix DMP-2831 material printer. Inject each of the six inks prepared into a separate printing cartridge.
2. Prepare an A4 size sheet of white copy paper.
3. Inkjet deposit circle-shaped spots of dye-encapsulated polymer nanoparticles onto the copy paper in 20 printing cycles (refer to Fig. 2 for the design of a triplicate sensor array and **Note 2** for setting of the printing conditions).
4. Dry the printed paper-based sensor array in a desiccator under vacuum for 2 days to remove the water and the ethylene glycol viscosity modifier.

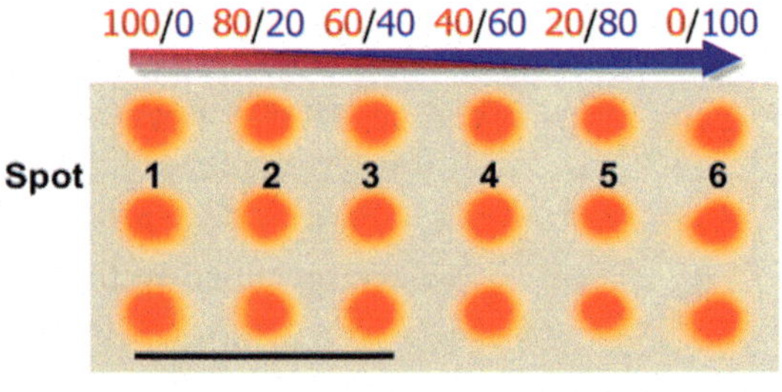

Fig. 2 Scanned image of an as-fabricated triplicate paper-based sensor array before sample exposure; the black scale bar represents a length of 5 mm; the numbers on top of the figure indicate the mixing ratio of the two types of dye-encapsulating polymer nanoparticle emulsions [red: pBzMA with lower polarity and blue: p(DEGMMA-co-MMA) with high polarity]. Reprinted with permission from Soga T, Jimbo Y, Suzuki K, Citterio D, Anal. Chem., 2013, 85, 8973-8978 Copyright (2013) American Chemical Society (**6**)

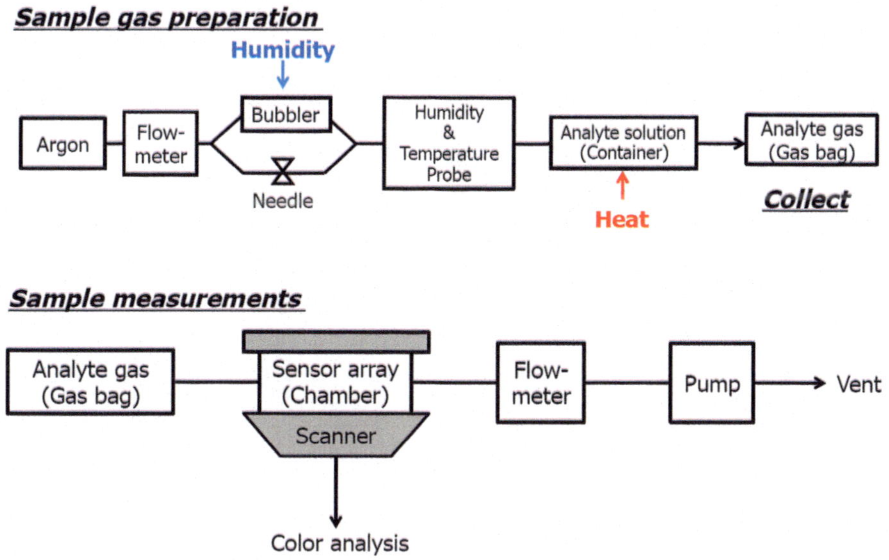

Fig. 3 Schematic illustration of sample gas preparation (top) and of sample gas measurement with the paper-based colorimetric sensor array (bottom). The sensor array is placed into a custom-made flow-through chamber with a transparent window facing the color scanner. This setup allows for the recording of color scans during gas exposure. Reprinted with permission from Soga T, Jimbo Y, Suzuki K, Citterio D, Anal. Chem., 2013, 85, 8973-8978 Copyright (2013) American Chemical Society (**6**)

3.5 Colorimetric Discrimination of Primary Aliphatic Amines

3.5.1 Sample Preparation and Measurement with Printed Paper-Based Sensor Arrays

A schematic illustration of the sample gas preparation and of the sample measurement with the inkjet-printed paper-based sensor array is shown in Fig. 3.

1. Separately generate vapors of seven primary aliphatic amines (*n*-methylamine, *n*-ethylamine, *n*-propylamine, *n*-butylamine, *n*-pentylamine, *n*-hexylamine, and *n*-heptylamine) by passing argon gas at a rate of 2 L/min through a bubbler with water and a sealed container with the respective amine in liquid form, which is heated with a heat gun (*see* Fig. 3, top, and **Note 3**).
2. Collect the generated amine gas in a gas-sampling bag.
3. Place a paper-based sensor array with the printed side facing down into a flow-through measurement chamber having a transparent bottom window and connect it to the gas-sampling bag and a pump (*see* Fig. 3, bottom).
4. Install the flow-through chamber on a Canon CanoScan 5600F color scanner and record a scan of the sensor array before amine exposure (settings: brightness 20, contrast 50), affording JPEG images (600 dpi).
5. Set the pump to a gas flow rate of 200 mL/min, and maintain the measurement chamber at ambient condition (25 ± 2 °C) for 20 min (*see* **Note 4**).
6. Scan the sensor array again to record an image after amine exposure.

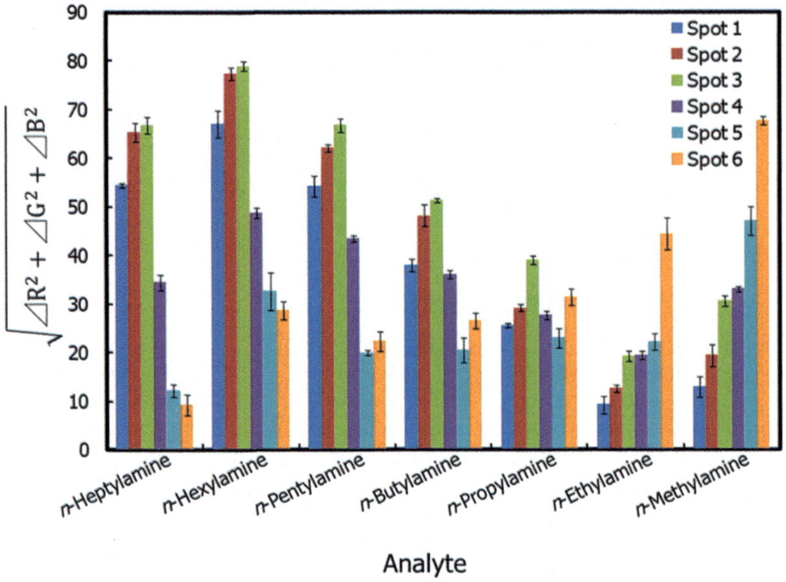

Fig. 4 Total color differences ($\sqrt[2]{\Delta R^2 + \Delta G^2 + \Delta B^2}$) observed after exposure of paper-based sensor arrays to 500 ppm of seven primary alkylamine gases at 50% relative humidity; error bars represent standard deviations of triplicate measurements; the composition of the sensing spots number 1–6 is as shown in Fig. 2, with polarity increasing from lower to higher numbers. Reprinted with permission from Soga T, Jimbo Y, Suzuki K, Citterio D, Anal. Chem., 2013, 85, 8973-8978 Copyright (2013) American Chemical Society (**6**)

3.5.2 Image Conversion and Data Processing

1. From the center of every spot in the array, extract the digital color information as numerical red, green, and blue (R, G, and B) color intensity values using the ImageJ (NIH, Bethesda, MD) software.
2. Calculate the differences of the R, G, and B values (ΔR, ΔB, ΔG) of every spot before and after exposure (RGB values after exposure—RGB values before exposure) to a gas sample (*see* Fig. 4 for an example of a bar graph showing analyte-dependent total color differences for every sensor spot in the array).
3. For improved graphical visualization of color change patterns, expand the resulting ΔR, ΔB, and ΔG values to the full 0–255 range, taking into consideration the minimum and maximum color difference values (*see* Fig. 5 for the resulting color change patterns of the sensor arrays upon exposure to seven amine samples represented as the ΔR, ΔB, and ΔG values expanded to the 0–255 range).
4. Apply principal component analysis (PCA) to the raw experimental data (non-expanded ΔR, ΔB, and ΔG values) as one method of multivariate statistical analysis allowing the

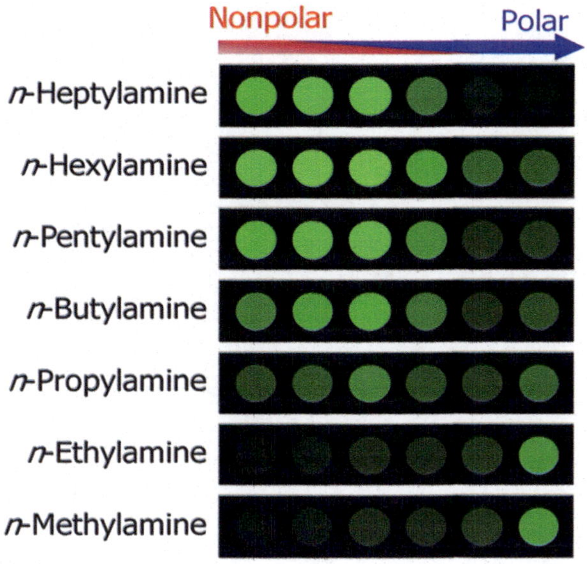

Fig. 5 Color difference patterns of paper-based sensor arrays after exposure to seven primary alkylamine gases. The patterns are represented as color differences (ΔR, ΔB, ΔG values) expanded to the full 0–255 range; analyte concentration: 500 ppm; relative humidity: 50%; the composition of the sensing spots is as shown in Fig. 2, with the leftmost spot corresponding to spot number 1. Reprinted with permission from Soga T, Jimbo Y, Suzuki K, Citterio D, Anal. Chem., 2013, 85, 8973-8978 Copyright (2013) American Chemical Society (**6**)

classification of measured samples (*see* Fig. 6 for a PCA plot obtained with the paper-based sensor arrays exposed to amine gases).

5. Perform agglomerative hierarchical clustering (AHC) analysis to the raw experimental data (non-expanded ΔR, ΔB, and ΔG values) to examine the multivariate distances between analyte responses in the sensor array (*see* Fig. 7 for an AHC dendrogram obtained with the paper-based sensor arrays exposed to amine gases).

3.6 Application to the Discrimination of Powdered Cheese Samples

It is known that natural cheeses generate various amines, whose distinct composition might be used as a characteristic fingerprint of the cheese type and origin. To demonstrate a simple practical application of the paper-based gas sensor array, it is applied to headspace analysis over two types of powdered cheeses: Parmigiano-Reggiano of controlled origin (*see* **Note 5**) and Kraft "100% Parmesan Cheese."

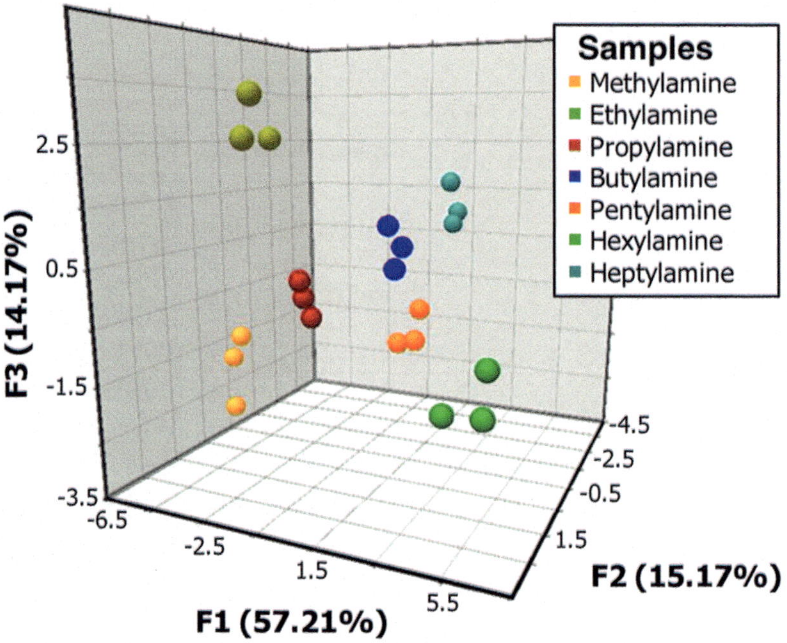

Fig. 6 Result of principal component analysis (PCA) for paper-based sensor arrays exposed to 500 ppm of seven primary alkylamines at 50% relative humidity; all experiments were performed in triplicate with three separate sensor arrays. Reprinted with permission from Soga T, Jimbo Y, Suzuki K, Citterio D, Anal. Chem., 2013, 85, 8973-8978 Copyright (2013) American Chemical Society (**6**)

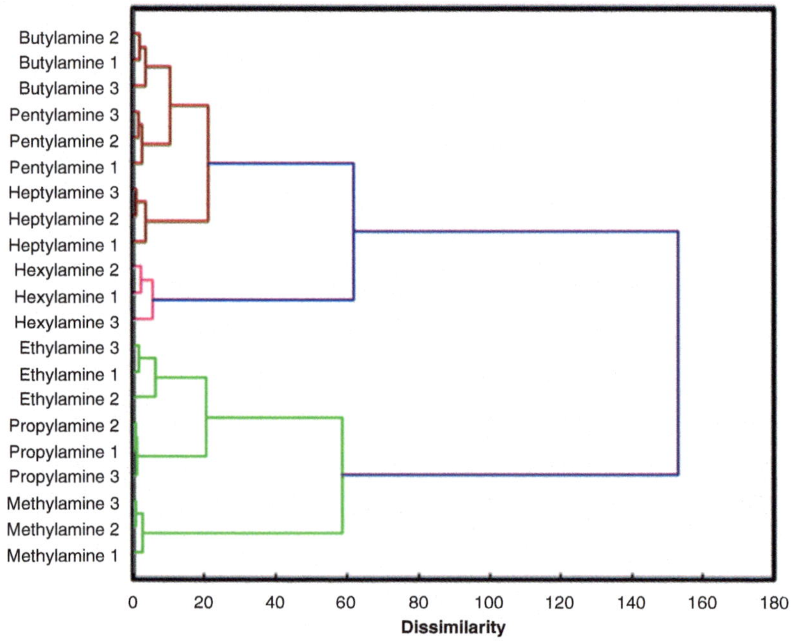

Fig. 7 Agglomerative hierarchical clustering (AHC) dendrogram for paper-based sensor arrays exposed to 500 ppm of seven primary alkylamines at 50% relative humidity; all experiments were performed in triplicate with three separate sensing arrays. Reprinted with permission from Soga T, Jimbo Y, Suzuki K, Citterio D, Anal. Chem., 2013, 85, 8973-8978 Copyright (2013) American Chemical Society (**6**)

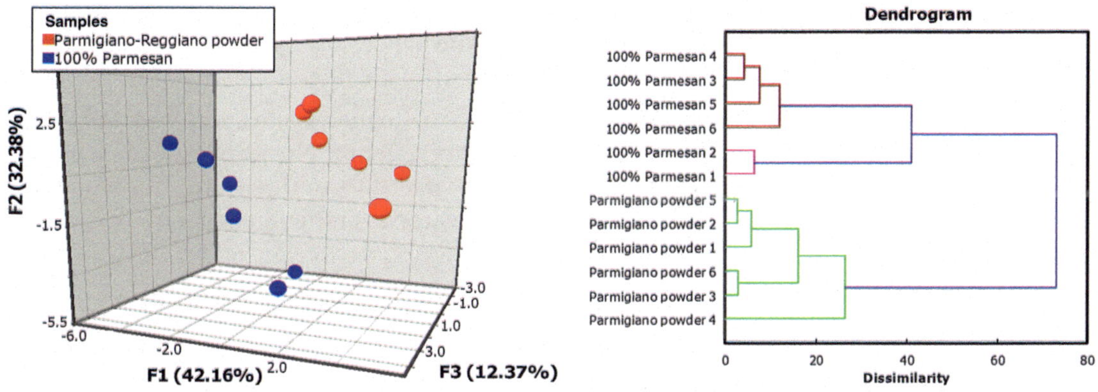

Fig. 8 (**a**) PCA plot and (**b**) AHC dendrogram for discriminating two powdered cheese samples (Parmigiano-Reggiano of controlled origin and "Kraft 100% Parmesan") with paper-based sensor arrays; all experiments were performed in six trials with three separate sensing arrays

3.7 Cheese Headspace Analysis with Prepared Paper-Based Gas Sensor Arrays

1. Attach a paper-based sensor array to the bottom side of a petri dish with a tape with the printed side facing downwards.
2. Capture the digital image of the prepared paper-based sensor array with the color scanner before exposure to the cheese sample (scanner settings: *see* **step 4** of Subheading 3.5.1).
3. Place 1.00 g of the powdered cheese sample around the fixed paper-based sensor array in the petri dish and close the lid of the petri dish.
4. Incubate for 30 min at ambient condition followed by capturing the digital image of the sample-exposed paper-based sensor array with the scanner.
5. Perform image conversion and data processing according to **steps 1–5** of Subheading 3.5.2. The obtained PCA and HCA plots are shown in Fig. 8.

4 Notes

1. The synthesized compound may contain traces of 1-bromohexane and *N*-hexylaniline, but can be used for the next reaction step without further purification.
2. To achieve a reasonably small size of the sensor array, a spot diameter of 1 mm with a spacing of 1 mm between spots is recommended. This size is sufficiently large to allow for the readout of the colorimetric response on the basis of color-scanned images. The material inkjet printer is set to a drop spacing of 20 μm during the printing process.
3. The amount of water is adjusted to generate samples with different relative humidity.

4. Shorter exposure times (*e.g.*, 2 or 10 min) also bring about colorimetric changes of the paper-based sensor array. However, visually observed colorimetric changes are faint.
5. The as-received Parmigiano-Reggiano cheese is cut into small blocks, which are frozen for 10 min. Next, the resulting blocks are milled to obtain a powdered form of cheese. The obtained powder is dried in a desiccator at ambient condition for several hours.

References

1. Askim JR, Mahmoudi M, Suslick KS (2013) Optical sensor arrays for chemical sensing: the optoelectronic nose. Chem Soc Rev 42 (22):8649–8682
2. Rakow NA, Suslick KS (2000) A colorimetric sensor array for odour visualization. Nature 406:710–713
3. Janzen MC, Ponder JB, Bailey DP, Ingison CK, Suslick KS (2006) Colorimetric sensor arrays for volatile organic compounds. Anal Chem 78(11):3591–3600
4. Bang JH, Lim SH, Park E, Suslick KS (2008) Chemically responsive nanoporous pigments: colorimetric sensor arrays and the identification of aliphatic amines. Langmuir 24 (22):13168–13172
5. Feng L, Musto CJ, Kemling JW, Lim SH, Suslick KS (2010) A colorimetric sensor array for identification of toxic gases below permissible exposure limits. Chem Commun 46 (12):2037–2039
6. Soga T, Jimbo Y, Suzuki K, Citterio D (2013) Inkjet-printed paper-based colorimetric sensor array for the discrimination of volatile primary amines. Anal Chem 85(19):8973–8978
7. Li J, Hou C, Huo D, Yang M, Fa H-b, Yang P (2014) Development of a colorimetric sensor Array for the discrimination of aldehydes. Sensors Actuators B Chem 196:10–17
8. Zhou X, Lee S, Xu Z, Yoon J (2015) Recent progress on the development of chemosensors for gases. Chem Rev 115(15):7944–8000
9. Ko HJ, Park TH (2016) Bioelectronic nose and its application to smell visualization. J Biol Eng 10(1):17
10. Chen Y, Fu G, Zilberman Y, Ruan W, Ameri SK, Zhang YS, Miller E, Sonkusale SR (2017) Low cost smart phone diagnostics for food using paper-based colorimetric sensor arrays. Food Control 82:227–232
11. López-Marzo AM, Merkoçi A (2016) Paper based sensors and assays: a success of the engineering design and the convergence of knowledge areas. Lab Chip 16(17):3150–3176
12. Yang Y, Noviana E, Nguyen MP, Geiss BJ, Dandy DS, Henry CS (2017) Paper-based microfluidic devices: emerging themes and applications. Anal Chem 89(1):71–91
13. Gong MM, Sinton D (2017) Turning the page: advancing paper-based microfluidics for broad diagnostic application. Chem Rev 117 (12):8447–8480
14. Yamada K, Shibata H, Suzuki K, Citterio D (2017) Toward practical application of paper-based microfluidics for medical diagnostics: state-of-the-art and challenges. Lab Chip 17 (7):1206–1249
15. Yamada K, Henares TG, Suzuki K, Citterio D (2015) Paper-based inkjet-printed microfluidic analytical devices. Angew Chem Int Ed 54 (18):5294–5310
16. Mohr GJ, Demuth C, Spichiger-Keller UE (1998) Application of chromogenic and fluorogenic reactands in the optical sensing of dissolved aliphatic amines. Anal Chem 70 (18):3868–3873
17. Mohr G, Spichiger U (1999) Reversible chemical reactions as the basis for optical sensors used to detect amines, alcohols and humidity. J Mater Chem 9(9):2259–2264
18. Mohr GJ, Nezel T, Spichiger-Keller UE (2000) Effect of the polymer matrix on the response of optical sensors for dissolved aliphatic amines based on the chromoreactand ETH T 4001. Anal Chim Acta 414(1):181–187
19. Korent ŠM, Lobnik A, Mohr GJ (2007) Sol-gel-based optical sensor for the detection of aqueous amines. Anal Bioanal Chem 387 (8):2863–2870
20. Comes M, Marcos MD, Martínez-Máñez R, Sancenón F, Villaescusa LA, Graefe A, Mohr GJ (2008) Hybrid functionalised mesoporous silica–polymer composites for enhanced analyte monitoring using optical sensors. J Mater Chem 18(47):5815–5823
21. Borisov SM, Mayr T, Klimant I (2008) Poly (styrene-block-vinylpyrrolidone) beads as a versatile material for simple fabrication of optical nanosensors. Anal Chem 80(3):573–582

Chapter 10

Label-Free Nanoplasmonic Biosensing of Cancer Biomarkers for Clinical Diagnosis

Alejandro Portela, Enelia C. Peláez, Olalla Calvo-Lozano, Mari C. Estévez, and Laura M. Lechuga

Abstract

Biosensing of cancer biomarkers enabling early diagnosis of cancer constitutes an essential tool for clinical intervention and application of novel therapies against cancer disease. Optical biosensor instruments as point-of-care (POC) devices and operating under label-free scheme have demonstrated to provide fast, simple, and high-sensitivity assays even at home care environment. Nanoplasmonic biosensors are thought to be a powerful tool for detection of complex analytes of relevant clinical applications. Using high-throughput fabrication techniques, large surface patterned with gold nanodisk structures is obtained showing surface sensitivities with limit of detection (LOD) in the order of picomolar concentration range. Here, we describe two major assay methodologies used for detection of lung and colorectal cancer, respectively. Particularly, we have selected a complementary hybridization DNA/RNA assay for the assessment of two miRNAs (miRNA-210 and miRNA-205) for detection of lung cancer. However, for colorectal cancer we present the detection of four tumor-associated antigen (TAA) biomarkers (MAP-KAPK3, PIM-1, STK4, and GTF2B) as possible TAA targets for autoantibody production. Strategies for detecting these biomarkers in real samples such as serum are also presented, demonstrating the capabilities of these assays to be transferred to real clinical settings.

Key words Nanoplasmonic biosensor, Cancer biomarkers, Immunoassay, Hybridization, microRNA, Autoantibodies, Colorectal cancer, Lung cancer, Clinical diagnosis

1 Introduction

New biosensing methodologies for the detection of cancer biomarkers enabling early diagnosis and intervention constitute an essential step to lead the battle against cancer disease. Modern biosensors, namely point-of-care (POC) devices, are overcoming limitations of classical analytical techniques allowing rapid, simple, and reliable assays even at the home care environment. Due to their high sensitivity under a label-free scheme, their immunity to electromagnetic (EM) interferences, and their stability in aggressive

Jessica E. Fitzgerald and Hicham Fenniri (eds.), *Biomimetic Sensing: Methods and Protocols*, Methods in Molecular Biology, vol. 2027, https://doi.org/10.1007/978-1-4939-9616-2_10,

environments, optical biosensors are top listed among the preferred choice for advanced POC sensing platforms.

Particularly, plasmonic biosensors are based on the resonant coupling of optical waves with the free electron cloud of noble metals, which produces an electromagnetic (EM) mode bound to the interface of the metal-dielectric. These clouds are commonly referred to as surface plasmons (SPs) and there are two classifications: (1) surface plasmon polaritons (SPP) or surface plasmon resonance (SPR) excited on thin metal films, and (2) the localized surface plasmon resonance (LSPR) created when the resonance mode is excited on sub-wavelength-sized metallic nanoparticles. SPR biosensors have been extensively studied and employed, due to their high sensitivity and simplicity. More recently, nanoplasmonic biosensors exploiting the LSPR have shown greater potential due to their miniaturization capabilities, simpler optical configuration, and higher surface sensitivity compared to its precursor SPR. Nanoplasmonic biosensors have moved from the fundamental stages of development focused on novel nanostructure geometries, their optical properties, and improvement of nanofabrication processes toward more relevant and real clinical and bioanalytical applications. However, only a limited number of research groups have succeeded in developing strategies that improve the performance, throughput, and/or integration of these label-free biosensors for conducting complex clinical assays associated with the early detection of cancer.

Lung cancer and colorectal cancer are the most common types of cancers related to morbidity and mortality worldwide [1]. Several biomarkers have been described as candidates for early detection of cancer; particularly relevant are those found at different levels of the genetic expression and regulation pathways: at DNA level (mutations), RNA level (alternative splicing variants and microRNA) [2, 3], and epigenetic level (altered methylation patterns) [4] and in proteins associated with tumors. Interestingly, those biomarkers can be found inside cancer malignant cells and in circulating body fluids such as serum, plasma, urine, and whole blood, and therefore can be analyzed for the detection of these diseases. In this chapter, optimized label-free biomarker detection for these two relevant cancers (lung and colorectal) is extensively described, including the materials and methods employed, from the fabrication of the nanoplasmonic sensing structure to the final biosensing assays.

In lung cancer disease, several microRNAs (miRNAs), endogenous small noncoding RNA molecules (18–22 nucleotides), have been found to be effective biomarkers for diagnosis and disease prognosis. They have a role as posttranscriptional regulators, by means of gene silencing, leading to the regulation of important signaling pathways involved in biological processes such as development, cell differentiation, and proliferation. They have gained relevance as biomarkers for noninvasive diagnosis of multiple pathologies such as cancer and neurodegenerative and cardiovascular diseases [5]. The evaluation of the miRNAs' differential expression levels can be a powerful tool for diagnosis, tumor-state differentiation, and prognosis [6]. More recently, several studies have explored the role of miRNAs in cancer, identifying some of them as effective biomarkers for diagnosis and/or prognosis (i.e., miRNA-210, miRNA-21, or miRNA-181). Moreover, as reported previously [7], the profile of the expression level of miRNAs is specific to certain cancer tissues supporting cancer classification. Particularly for the biosensing of lung cancer, we have selected a complementary hybridization DNA/RNA assay for the assessment of two miRNAs (miRNA-210 and miRNA-205) widely reported in the literature [8, 9].

As for colorectal cancer (CRC), immunogenic proteins known as tumor-associated antigens (TAA) that are related to malignant CRC cell growth have shown their potential as biomarkers. TAA stimulates cellular and humoral immune response, inducing autoantibody production. Although the autoantibodies' regulatory mechanism is not completely understood, they are considered promising biomarkers for the early diagnosis of numerous cancer types due to their high stability and large production even with a minimal quantity of tumor antigen [10]. Over the last few years, a number of studies have been directed toward defining specific TAAs capable of detecting TAA autoantibodies with high accuracy and reliability [11]. We have selected four TAA biomarkers (MAPKAPK3, PIM-1, STK4, and GTF2B) as possible TAA targets for autoantibody production in colorectal cancer [12]. The presence of the autoantibodies associated with TAA proteins can be detectable at a very early stage in tumor development, even before adenoma formation, which reinforces their potential as biomarker candidates for improving CRC diagnosis at the earliest stages.

Several advantages have been endorsed to the nanoplasmonic biosensors, one of which is the short decay length of the EM field (tailored by shape, size, and composition of the nanostructure) which has a range in the same order of the typical size of targeted biomolecules [13]. This condition leads to an increase in sensitivity of small changes in the refractive index (RI) of the nanostructures' environment, since the analytes can cover a larger fraction of the EM field sensing probe. The large confinement of the EM field surrounding the nanostructure provides sufficient sensitivity for direct and label-free detection of small biomolecules, even to a single-molecule level. Finally, the ease of matching the conditions needed to excite the plasmon resonance simplifies the optical instrumentation, facilitating the miniaturization and implementation as POC devices.

The excitation of LSPR in a metallic nanoparticle by a light wave is illustrated in Fig. 1. The variation of the RI in the close metal-environment interface strongly modifies the resonance condition.

Fig. 1 LSPRs are excited on metallic nanostructures with an added capturing layer enabling the direct and specific recognition of analytes. The triangles (green) represent the specific antigen binding to the capturing antibody attached on the gold nanostructures. The circles (red) represent nonspecific antigens to the capturing antibodies, which do not react with these antibodies

The response of the resonance condition in terms of wavelength shift response $\Delta\lambda_{\mathrm{LSPR}}$ can be described through Eq. 1 [14]:

$$\Delta\lambda_{\mathrm{LSPR}} = m\,\Delta n\left(1 - e^{-2d/l_d}\right) \tag{1}$$

where $\Delta\lambda_{\mathrm{LSPR}}$ is the wavelength shift response, m is the refractive index sensitivity, Δn is the change in refractive index induced by the adsorbate, d is the effective adsorbate layer thickness, and l_d is the characteristic electromagnetic field decay length. A capturing layer properly attached on the metallic features located over a planar substrate enables the detection of specific binding events as illustrated in Fig. 1.

Local RI changes, such as those induced by the selective attachment of biomolecules at the surface of the nanostructures, can strongly modify the excitation conditions of the LSPR, and the spectra are monitored over time for tracking the shift of the LSPR extinction peak (Fig. 2a, b). In the following sections, several bioassays using this refractometric sensing approach based on the real-time monitoring of the LSPR resonance peak displacement are presented.

2 Materials

2.1 Equipment

1. UV/Ozone cleaner, ProCleaner™ Plus (BioForce Nanoscience, USA).
2. Refractometer J57 RUDOLPH (Hackettstown, NJ, USA).

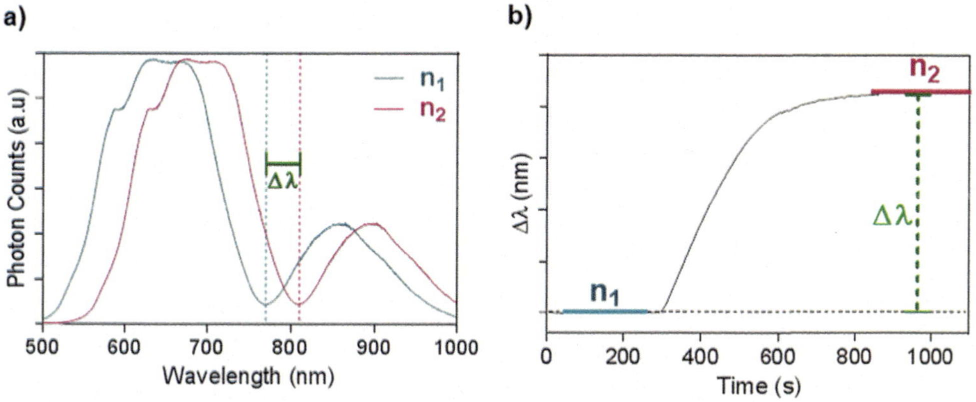

Fig. 2 (**a**) Spectra showing wavelength displacement of the resonance peak ($\Delta\lambda$) caused by the refractive index changes. (**b**) Real-time monitoring of wavelength shift ($\Delta\lambda$) vs. time (s)

3. Plasma cleaner reactor: Standard plasma system Femto, version A (40 kHz, 100 W), from Diener Electronic GmbH (Ebhausen, Germany).
4. Ultrasonic bath (FB15047, FisherBrand, Germany).
5. Reactive ion etching is done in a clean room facility (class 100).
6. Thermal mixer (Eppendorf AG22331, Hamburg, Germany).

2.2 Setup Components

1. Nanoplasmonic chips fabricated on glass surface (No. 4, 22 × 22 mm^2, Thermo Scientific Menzel-Glazer, Germany) through the hole-mask colloidal lithography (HCL) process.
2. Custom-made fixed-angle stage for adjusting the incident illumination angle at 80°.
3. A trapezoidal prism of 80° (K9, $n = 1.52$) is used with matching oil ($n = 1.52$).
4. Custom-made flow cell (volume = 4 μL) made of Teflon that ensures a controlled and constant flow over the sensing area.
5. Syringe pump NE-1000 (New Era Pump Systems, USA) for maintaining a constant flow in the sensing area.
6. A manual injection valve (IDEX Health and Science, V-451, USA) allows the injection of samples or solutions through the flow cell.
7. A halogen light (HL-2000, Ocean Optics, USA) is fibered to the stage, collimated (C330TMEB, Thorlabs, Germany), and configured in transverse electric (TE) polarization using linear polarizer LPVIS050 (Thorlabs, Germany) mode for the LSPR excitation.

8. A compact charge-coupled device (CCD) spectrometer (Ocean Optics, Jazz Module, USA) connected through a fiber is used to analyze the reflected light.
9. Reflectivity spectra are acquired every 3 ms and 300 consecutive spectra to average to provide the spectrum to be analyzed. Real-time changes in the resonance peak (λ_{LSPR}) are tracked in real time via polynomial fit using homemade readout software (National Instruments, LabVIEW, USA).

2.3 Reagents

1. Organic solvents: Acetone, ethanol 96%, toluene, ethanol absolute, 2-propanol.
2. Hydrogen peroxide 30% (H_2O_2), sulfate latex beads 8 wt%, 0.1 μm, polymethyl methacrylate 950 K (PMMA), anisole, and poly(diallyldimethylammonium) chloride (PDDA solution 20 wt%, Mw 400–500 K).
3. 1-Ethyl-3(3-dimethylaminopropyl)carbodiimide hydrochloride (EDC), N-hydroxysulfosuccinimide sodium salt (NHS), ethanolamine hydrochloride, Tween 20, bovine serum albumin (BSA), sodium phosphate monobasic (NaH_2PO_4), sodium phosphate dibasic (Na_2HPO_4), sodium chloride (NaCl), sodium hydroxide (NaOH), trisodium citrate dihydrate (SSC), diethyl pyrocarbonate (DEPC), ethylenediaminetetraacetic acid (EDTA), 4-(2-hydroxyethyl)piperazine-1-ehanesulfonic acid (HEPES), potassium chloride (KCl), N-morpholinoethanesulfonic acid (MES), potassium dihydrogen phosphate (KH_2PO_4), hydrochloric acid (HCl), sulfuric acid 96% (H_2SO_4).
4. Isobutyl(trimethoxy)silane.
5. Sulfate Latex Beads S37204 with diameter of ~100 nm of polystyrene (PS) (Invitrogen, Thermo Fisher Scientific, USA).
6. Alkanethiol 16-mercaptohexadecanoid acid (MHDA).
7. Thiol-poly(ethylene)glycol-amine ($SH\text{-}PEG\text{-}NH_2$, MW ~3.400 g mol^{-1}), thiol-poly(ethylene)glycol-carboxyl (SH-PEG-COOH, MW ~2.000 g mol^{-1}), methoxyl-poly(ethylene)glycol-silane (m-PEG-silane, MW ~2.000 g mol^{-1}), poly-L-lysine-graft-PEG (PLL-g-PEG, MW ~70,000 g mol^{-1}).
8. DNA capture probes incorporating thiol group (SH-) at the 5′-end and miRNA nucleotide sequence (Table 1).
9. Antibody anti-DNA-RNA hybrid [S9.6].
10. Recombinant proteins TAAs (GT2b (general transcription factor IIB), MAPKAPK3, PIM-1, STK4, EDIL-3) were produced in bacterial expression systems and purified according to previous studies [15, 16]. Antibodies anti-GTF2b (D-3 (Sc-225) and C-18 (sc271736)) and anti-PIM-1 (Ab117518MAb) were purchased from Santa Cruz Biotechnology (USA); antibodies

Table 1
DNA capture probes and miRNA nucleotide sequences for miRNA detection

	miRNA Sequence	Capture probe
Target biomarker miRNA-210-3p	5′CUGUGCGUGUGA CAGCGGCUGA 3′	5′SH-T15-AGCCGCTGTCACACGCACA3′
Reference miRNA-205-5p	5′UCCUUCAUUCCACC GGAGUCUG 3′	

anti-MAPKAPK3 (3F4MAb), anti-STK4 (1D7-8A10MAb), and anti-EDIL-3 (Ab67573 PAb) were purchased from Abnova (Germany).

11. DEPC-H_2O (Milli-Q water (deionized water from a Milli-DI® Water Purification System, Merck Millipore, USA)) incubated overnight with 0.1% DEPC and autoclaved at 121 °C during 1 h. All solid plastic and glass materials were sterile (*see* **Note 1**).
12. Bond-Breaker™ TCEP Solution (Tris(2-carboxyethyl) phosphine hydrochloride solution).

2.4 Buffer Composition

1. Phosphate buffer saline 50 mM (PBS-5X): Solution of 750 mM NaCl, 33 mM Na_2HPO_4, 17 mM NaH_2PO_4, and 2 mM EDTA in DEPC-H_2O water, pH adjusted to 7. Store at room temperature (RT). We recommend preparing a stock solution (200 mM PBS pH 7–7.5).
2. PBS-1X: Solution of 1.37 M NaCl, 0.027 M KCl, 0.015 M KH_2PO_4, 0.080 M Na_2HPO_4, pH 7.5.
3. PBST-0.5 (PBS-1X, with 0.5% Tween 20, pH 7.5).
4. Sodium citrate buffer saline (SSC-5X): Solution of 75 mM SSC, 750 mM NaCl, 4 mM EDTA in DEPC-H_2O water, pH adjusted to 7. Store at RT. We recommend preparing a stock solution (20 × SSC pH 7).
5. MES buffer: Solution of 100 mM MES, 500 mM NaCl pH 5.5 in Milli-Q water.
6. Solution of ethanolamine 1 M pH 8.5 in Milli-Q water.
7. HEPES buffer: Solution of 10 mM HEPES, 150 mM NaCl, pH 7.4 in Milli-Q water.

3 Methods

One of the initial processes for developing a sensitive plasmonic biosensor is the numerical modeling of the plasmonic nanostructures to be employed. This theoretical step seeks to obtain a higher

EM field intensity enhancement, confinement, and sensitivity to RI changes. Several parameters of the nanostructures can be simulated to optimize their response—namely, composition, shape, thickness, and other geometrical parameters, as well as the excitation and light collection configuration.

Several geometrical structures and complex architectures requiring either sophisticated fabrication methods or costly optical configurations have been extensively reported in the literature [17]. One of the main constraints limiting the realization of practical devices based on nanoplasmonic biosensors is obtaining plasmonic nanostructures of high sensitivity that can be fabricated with simple, scalable, and cost-effective methods such as mask-colloidal lithography, nanoimprinting lithography, and soft lithography for nanoscale patterning. In this regard, mask colloidal lithography is a feasible option for production of large-area sensor chips in a laboratory environment (with almost no requirement of a clean room facility), which contain a randomly distributed array of nanostructures, namely nanodisks, supporting an EM waveguide mode of extraordinary sensitivity to RI fluctuations [18].

For the characterization of the sensor chips, a customized optical setup is required to excite the in-plane polarization of the LSPR of the nanodisks (onto a planar glass substrate) under total internal reflection (TIR) conditions. The in- and out-coupling of the light is configured through optical fibers, and the collimated incident beam is linearly polarized (TE) for the LSPR excitation, whereas the reflected light is processed via an external CCD spectrometer. Two additional external components, a broadband light source and a flow control system, are also employed.

The biosensing applications of the nanoplasmonic sensors require a careful selection of the conditions for immobilizing a biorecognition (or capturing) element on the plasmonic sensor surface. This biofunctional layer defines the selectivity of the sensor toward the detection of specific analytes in complex biological samples. Two distinctive examples of nanoplasmonic biosensing are discussed in detail in the following sections. One is based on an immunodetection assay (antigen–antibody-specific interaction), and the second one, on a complementary hybridization DNA/RNA assay.

3.1 Numerical Analysis of Nanoplasmonic Structures

For a better understanding of the EM field on the surface of the plasmonic nanostructures, their resonance modes, enhancement, and spectral location can be numerically approximated. The 3D finite-difference time-domain method (FDTD) is a theoretical calculation method that solves the Maxwell equations for complex geometries, whose optical properties lack analytical solutions. Using the FDTD from *FDTD solutions* (Lumerical Inc., Canada) software it is possible to obtain the spectra of the Au nanodisks excited with a broadband illumination source.

1. Choose experimental dielectric constants (i.e., Johnson and Christy [19] for Au) and the glass substrate (SiO_2) with refractive index $\eta_g = 1.507$ for the planar substrate supporting the nanostructure.
2. Define the aspect ratio 5:1 (diameter/height) of the nanodisks in order to set the resonance peak in the visible range where the CCD spectrometer has an adequate response and the best figure of merit (FOM) of the resonance peak given by the full-width-half-maximum (FWHM) of the spectral curve. In agreement with previous theoretical and experimental reports of our group, a diameter of 100 nm and thickness of 20 nm resulted in the optimum dimensions [20]. The environment of the nanodisk was assumed to be water $\eta_{env} = 1.328$. Choose a plane wave illumination source incident from below with a normal incident angle and the light beam linearly polarized parallel (in-plane) to the sensor substrate, choosing a wavelength range between 0.5 and 1.0 μm.
3. Set a variable size rectangular mesh of 2 nm in the vicinity (40 nm) of the surface; boundary conditions as perfectly matched layer (PML) can be used for determining the field intensity distribution and the LSPR spectral response of the nanodisks.
4. Define the monitors of interest in order to represent the near-field profile around the nanostructures. The extinction feature of the nanodisks at the far field can be obtained from these monitors.
5. It can be verified that calculated optical response shows a strong dependency on structural dimensions. Particularly, the resonant peak position quasi-linearly increases with an increase of the length of the nanodisks, as shown in Fig. 3a, which evidences their property of spectral tunability.

   Similar properties can also be achieved for nonsymmetric nanodisks, particularly elliptical nanodisks, without requiring fabrication of different nanodisk sizes. Owing to the anisotropy of the structure, it presents two characteristic peaks produced by plasmonic resonances (dipole modes) which are located within the visible (VIS) and near-infrared (NIR) spectral region. Therefore, besides the spectral tunability with respect to the overall size of the structure, it also enables the tunability of the resonance wavelength by modifying the orientation of the linearly polarized light with respect to the longer axis radio of the ellipsoid, as shown in Fig. 3b, which was obtained with similar conditions for the numerical simulations except for the nanodisks' dimensions.

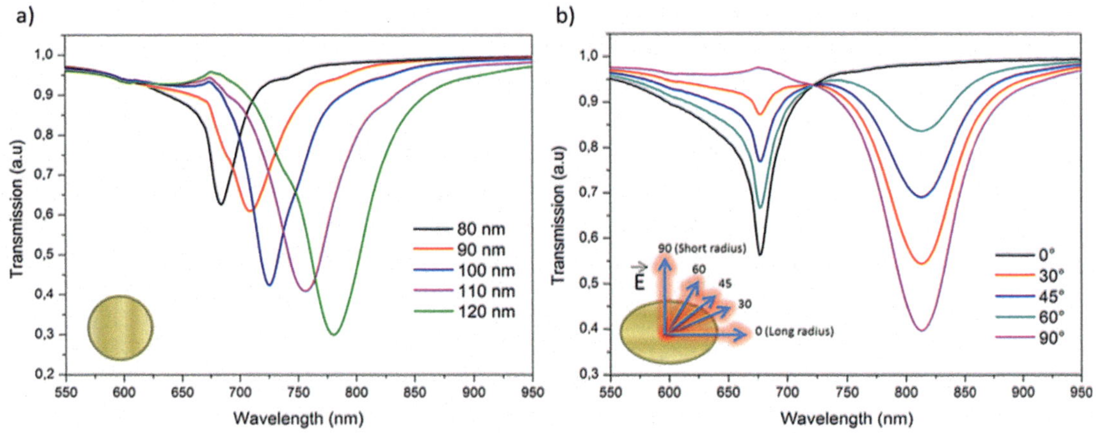

Fig. 3 (**a**) Transmission spectra of round nanodisks with size ranging from 80 to 120 nm in water environment excited with an in-plane linearly polarized light. (**b**) Transmission spectra of elliptical nanodisks (120 × 80 nm), excited with an in-plane linearly polarized light with different orientations (0°–90°) with respect to the longer axis of the ellipsoid

3.2 Nanoplasmonic Chip Fabrication

Nanoplasmonic chips of random arrays of gold nanodisks (diameter = 100 nm, height = 20 nm, surface density = 6–7%) are fabricated based on initial protocol of hole-mask colloidal lithography (HCL) [21]. Figure 4 summarizes the nanofabrication process that is carried out following a modified protocol as described below. Standard cleaning process (not shown) of the glass slides is performed using SDS, HCl, and Milli-Q water prior to drying with a flow of N_2. Sonication of the chips for 5 min at 50 °C, first in acetone and then in isopropanol, is followed by drying under N_2 flow.

1. Immerse the chip overnight in a solution of 10 μL of isobutyl (trimethoxy)silane in 5 mL toluene, to improve the adhesion of a PMMA layer. Rinse the chips using toluene followed by isopropanol and dry under N_2 flow.
2. Obtain a deposition of a PMMA (4% in anisole) with a thickness of approximately 210 nm using a spin coating process (4000 rpm, 1500 r s^{-2}). Immediately, bake the sample at 165 °C for 5 min.
3. Expose the chips to a short O_2 plasma treatment (10 s, 75 W, 75 mTorr) to increase the hydrophilicity of the PMMA layer.
4. Deposit a cationic solution of PDDA (0.2 wt%) suspended in water dropwise onto the PMMA layer and incubate for 1 min. Rinse the sample with water and dry under N_2 flow.
5. Dilute the polystyrene beads (PS) in water (0.01 wt%) and sonicate for 10 min prior to drop-coating (1 min) on the chip. Rinse the chip and immerse in water (95 °C, 3 min) to fix the PS beads on the PMMA layer.

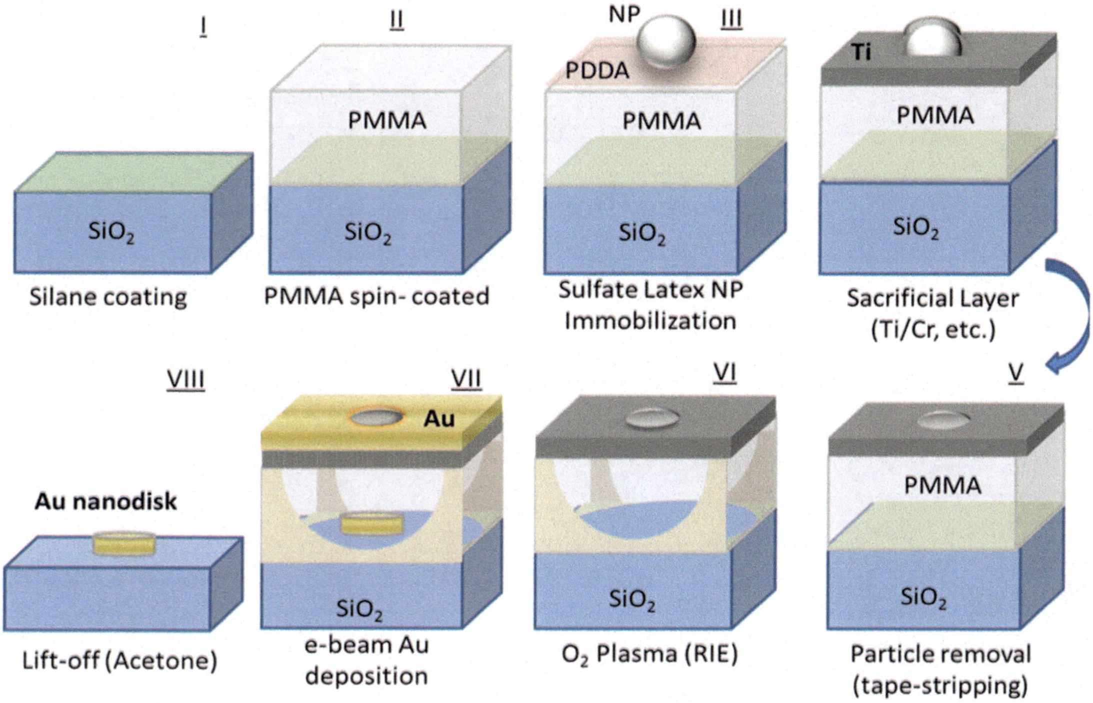

Fig. 4 Schematic representation of the nanofabrication process of gold nanodisks non-ordered arrays based on HCL

6. Deposit a 15 nm sacrificial layer of titanium (Ti) that can withstand an O_2 plasma treatment to create the hole-mask template.
7. Remove PS beads by a tape-stripping technique to create a perforated Ti hole-mask.
8. Selectively etch the exposed PMMA with a reactive ion etcher (RIE) process using oxygen plasma (5 min, 400 W, 75 mTorr, 50 sccm O_2-flux) to create a cavity underneath the hole left by the PS nanoparticle on the Ti layer.
9. Perform electron beam evaporation of a gold layer to form the plasmonic nanodisks by evaporating a thin 1 nm metal adhesion layer of Ti, followed by 19 nm of Au.
10. Expose Au nanodisk arrays using a liftoff process in acetone (thoroughly rinsing and immersing for 1 h) which removes the hole-mask.
11. Lastly, rinse the chip with isopropanol and dry with N_2 flow.

The process is flexible for variations introduced to create non-symmetric nanodisks. A slight modification of the HCL fabrication procedure described above leads to the creation of an array of elliptical nanodisks. In this case, the evaporation of the Ti sacrificial layer can be conducted at a rotation angle (γ) of the sample with

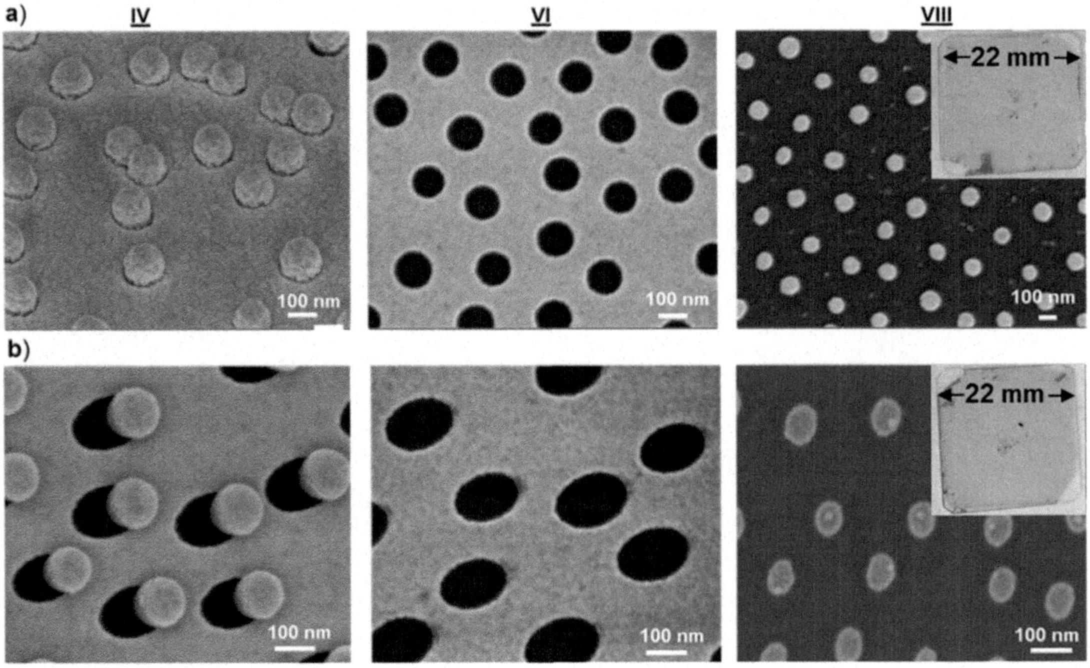

Fig. 5 Scanning electron microscopy (SEM) of round (**a**) and elliptical (**b**) nanodisks after key steps of the fabrication process. The images correspond to the condition of the chip after (IV) sacrificial layer deposited over PS particles, (VI) tape-stripping of PS particles, and (VII) liftoff in acetone. A scale bar (100 nm) is indicated for each image. The insets show a photograph and dimensions of the square chip obtained for each sensing nanostructure

respect to its vertical axis. The values of γ can be used as a variable to define the long axis length and to tune the spectral position of the LSPR. In our case an angle γ of 5° was employed, resulting in elliptical nanodisks with long and short radii of 90 nm and 80 nm, respectively. Scanning electron microscope (SEM) images of the fabricated round and elliptical nanodisks are depicted in Fig. 5a, b, respectively. The images illustrate the crucial steps of the fabrication process: after sacrificial layer deposited over PS particles (IV), tape-stripping of PS particles (VI), and liftoff in acetone (VII), the final sensing structures are ready for biofunctionalization. Large-scale sensor chips of 22×22 mm^{-2} are produced (as shown in the insets of Fig. 5) which are cut manually prior to using them in the optical setup.

3.3 Setup Configuration

The complete platform contains the optical setup based on reflectivity measurements in TIR configuration as shown in Fig. 6. Additional external components such as the light source, the flow pump control, and the spectrometer are also set conveniently. Besides these components, the system requires in-house software running on a PC to acquire, process, and visualize the data from the spectrometer, allowing real-time monitoring of the LSPR spectral shift.

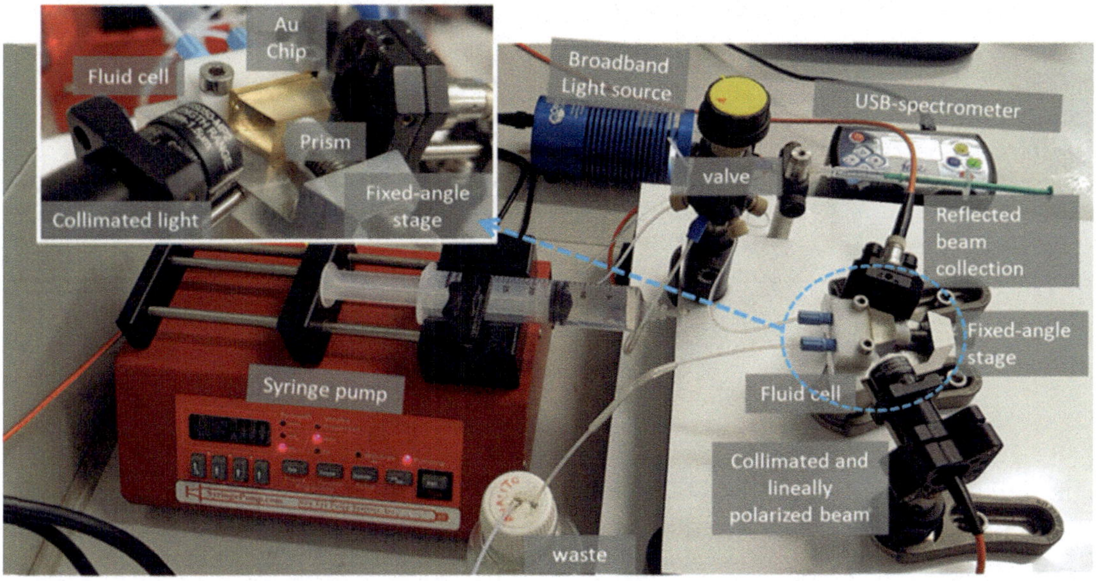

Fig. 6 Photograph of the LSPR sensor device showing the main optical sensing unit (inset) and the external components and connections required

1. Place the LSPR sensor chip onto a platform containing all the optical components to couple the light on the sensor chip and the fluidic components to deliver the liquid sample to the sensor chip (fluidic cell, tubing, and injection valve).
2. Connect the halogen lamp through an optical fiber to a fixed-angle stage containing a collimator, linear polarizer, and trapezoidal prism to excite the nanoplasmonic sensor chip using a conventional Kretschmann configuration. The broadband light must be linearly polarized (TE) and the collimated beam incident with optimal angle of 80°.
3. A fluidic cell should be mounted on the fixed-angle stage to clamp the nanoplasmonic sensor chip with the trapezoidal prism. The fluidic cell requires an inlet and outlet for liquid sample to flow.
4. Connect a syringe pump, functioning as flow delivery system, through a manual injection valve to the inlet of the fluidic cell. The loop of the injection valve has a volume of 200 μL to insert samples on the flow system. Connect the outlet of the fluidic cell to the waste reservoir.
5. Couple the reflected light using an optical fiber to a CCD spectrometer connected to a computer. Real-time processing of the spectra is required in order to run an algorithm to track the displacement of the spectrum and display the results of the experiment.

3.4 Evaluation of Bulk Sensitivity of Fabricated Nanoplasmonic Structures

In order to evaluate the response of the sensor, the bulk sensitivity of the LSPR sensor is performed. For that purpose, Milli-Q water is flowed on the sensor surface as running buffer.

1. Prepare solutions of different RI (e.g., different dilutions of PBS).
2. Determine the RI of the different solutions using a commercial refractometer and calculate the difference of the refractive index (Δn) between the Milli-Q water (running buffer) and the solutions with different RI.
3. Initiate a continuous flow of the running buffer (35 μL min^{-1}) into the sensor chip.
4. Inject the different RI-known solutions in turns, registering the changes of the spectra over time and tracking the shift of the LSPR extinction peak ($\Delta\lambda_{\text{LSPR}}$).
5. Plot the wavelength shifts ($\Delta\lambda_{\text{LSPR}}$) *versus* refractive index variation (Δn), obtaining a calibration curve. The slope of the fitting curve represents the bulk sensitivity (S_{bulk}), expressed in nm·RIU^{-1}. Figure 7 shows the sensor signal and the calibration curve using elliptical nanodisks (90 nm × 80 nm) for a TE-polarized incident light and oriented parallel to the short radius of the nanostructure.
6. Calculate the limit of detection (LOD), Δn_{min}, through Eq. 2:

$$\Delta n_{\text{min}} = \frac{\Delta\lambda_{\text{min}}}{S_{\text{bulk}}} = \frac{\Delta S_{\text{R,min}}}{S_{\text{bulk}}} = \frac{3\tilde{n}\sigma_{S_{\text{R}}}}{S_{\text{bulk}}} \tag{2}$$

where $S_{\text{R,min}}$ is the minimum sensor response, defined as three times the system noise, σ_{SR}. LOD for LSPR biosensors is commonly found in the range of 10^{-5}–10^{-6} RIU.

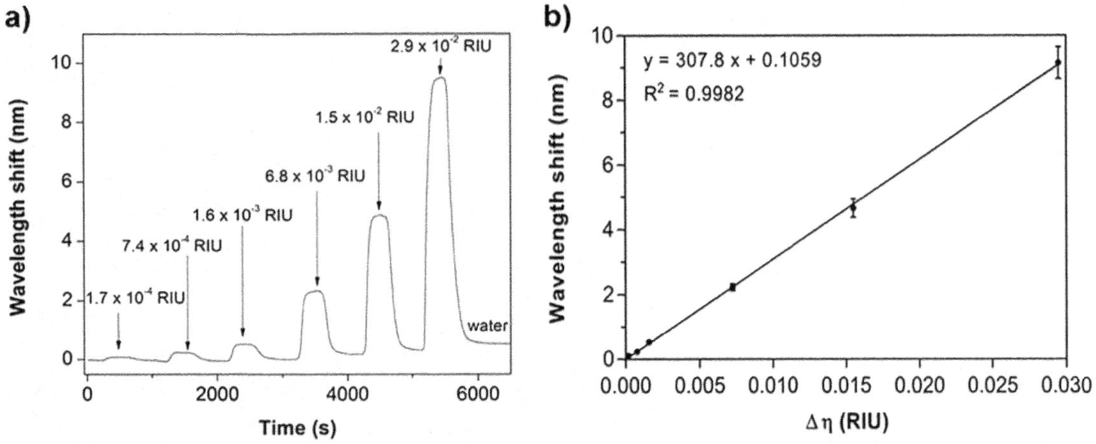

Fig. 7 Bulk sensitivity of elliptical nanodisks. (**a**) Sensor response to six PBS solutions with different refractive indexes. (**b**) Calibration curve showing the relationship between the variation in the refractive index (Δn) and the sensor response ($\Delta\lambda_{\text{LSPR}}$)

3.5 Surface Biofunctionalization for Hybridization DNA/RNA Assays

Biosensing application for miRNA detection is based on a complementary hybridization DNA/RNA assay as shown in Fig. 8. For the analysis of miRNA, the sensor chip surface is biofunctionalized with DNA capture probes that contain the complementary sequence to the target miRNA. The specific miRNA interacts with the immobilized probe and generates a change in the RI directly related to the miRNA concentration in the sample (Fig. 9).

The most applied strategy for chemical modifications of gold sensor surfaces is based on self-assembled monolayers (SAM) of thiols, owing to the strong affinity between gold and thiol groups [22]. The design of the DNA capture probes necessitates their functionalization with thiol groups in order to attach them to the

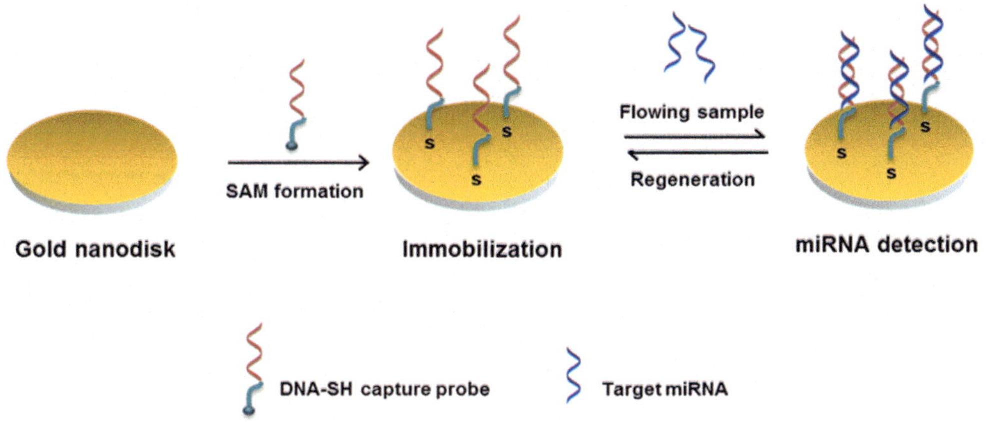

Fig. 8 Complementary hybridization assay for miRNA detection

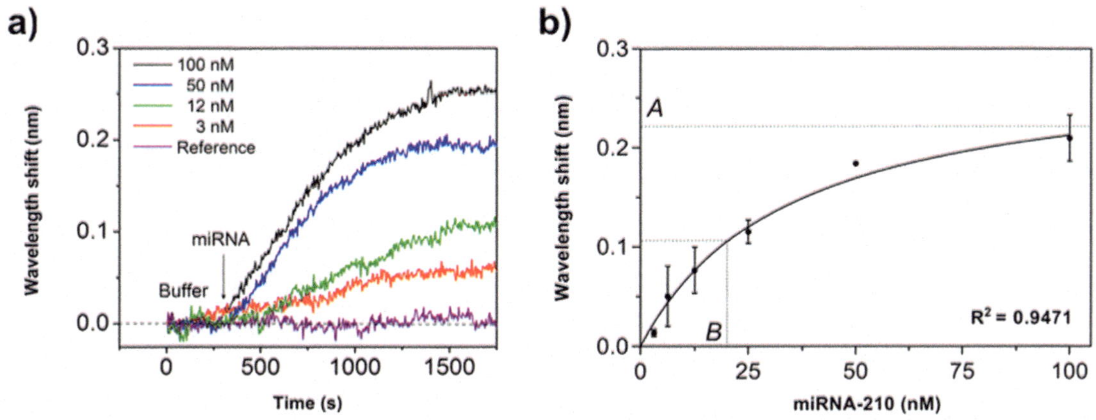

Fig. 9 miRNA-210 detection with LSPR biosensor. (**a**) Sensor response ($\Delta\lambda_{LSPR}$) to four samples with different miR-210 concentrations. The reference value corresponds to a noncomplementary miRNA (100 nM miRNA-205). (**b**) Calibration curve showing the relationship between the miRNA (miRNA-210) concentration and the sensor response ($\Delta\lambda_{LSPR}$)

sensor chip surface. Thus, a complementary hybridization assay is performed to conjugate each DNA capture probe with a thiol group that interacts with the gold surface and a spacer (sequence of 15 thymines) that keeps away the recognition DNA from the surface in order to facilitate the accessibility of the DNA complementary sequence to the target miRNA.

In order to avoid contamination of the surface and to enhance reproducibility, the LSPR chips must be cleaned prior to biofunctionalization. Once the LSPR chips are clean, DNA capture probes are immobilized in a single biofunctionalization step.

3.5.1 Cleaning Procedure

1. Rinse the sensor chip sequentially with acetone and 2-propanol and dry under N_2 stream.
2. Place the cleaned LSPR chip in an oxygen plasma cleaner (100 W, 45 sccm) for 2 min (*see* **Note 2**).
3. Rinse the sensor chip generously with ethanol 96% and 2-propanol and dry carefully under N_2 stream.

3.5.2 Immobilization of DNA Capture Probes

The cleaned sensor chips are placed on the experimental setup. The immobilization of the DNA capture probes is performed inside the measuring setup (in situ) at a constant flow rate.

1. Prepare a solution of DNA capture probes in 50 mM PBS buffer, and add 100 nM TCEP (TCEP is a reducing agent that avoids disulfide bonds between thiols of the DNA capture probes). Incubate the mixture for 20 min at 36 °C, 750 rpm, on the thermal mixer. DNA probe concentrations must be optimized for each application, although concentration typically ranges from 1 to 2 μM of DNA probes.
2. Initiate a continuous flow (18 μL min^{-1}) of DEPC-H_2O.
3. Inject the DNA capture probe mixture (300 μL) using the same flow rate until the sensor's signal stabilizes (the sensor response should achieve the baseline). The sensor chip is now functionalized and ready to be used for miRNA detection.

3.5.3 Detection by Complementary Hybridization Assay

1. Initiate a continuous flow (18 μL min^{-1}) of SSC 5×.
2. Prepare a series of solutions of different concentrations of miRNA-210 in SSC 5× (300 μL, standard solutions). Concentrations of the standard solutions depend on the LOD for each application. For miRNA-210, the standard solutions were in the nM to pM range in our studies.
3. Inject the miRNA-210 standard solutions at 18 μL min^{-1}. This step produces a wavelength shift ($\Delta\lambda_{LSPR}$) due to the interaction of the miRNA with the immobilized DNA capture probes.
4. Regenerate the sensor surface by injecting 5 mM NaOH solution (18 μL min^{-1}, 60 s) and wait until the signal stabilizes (*see*

Note 3). At this point, the sensor surface is ready for the measurement of a second standard solution. In Fig. 9a, the sensor response for five samples (in SSC 5×) is depicted, in which the absence of signal with a noncomplementary miRNA (reference) can be observed.

5. Construct a graph of the sensor response *versus* miRNA-210 concentration and fit the data to an appropriate curve-fitting model to obtain the calibration curve (*see* **Note 4**). LOD should be in the desired range for the application. If not, try to optimize both immobilization and detection conditions. The miRNA-210 usually appears at fM–nM range in human plasma, although for clinical applications an LOD of pM is adequate.
6. Quantify the concentration of miRNA in samples of unknown concentrations:
   (a) Inject the sample, wait until the signal stabilizes, and register the value of the sensor response ($\Delta\lambda_{LSPR}$).
   (b) Estimate the concentration of miRNA-210 in unknown samples by evaluating the value of the sensor response ($\Delta\lambda_{LSPR}$) in the calibration curve previously obtained.

3.5.4 Detection by Antibody Amplification

MiRNAs are small macromolecules (7000 Da). They can appear in low concentrations (aM–fM) in serum of real patient samples. In order to improve the sensitivity of the LSPR biosensor to detect extremely low concentrations of miRNA, an amplification step may be needed. Additionally, in order to reduce the nonspecific adsorptions and avoid background signals a *blocking* step of the sensor surface is required (*see* **Note 5**).

The most common strategy is employing an antibody that specifically interacts with the bound DNA/RNA hybrids (Fig. 10). The antibody does not cross-react with double-stranded

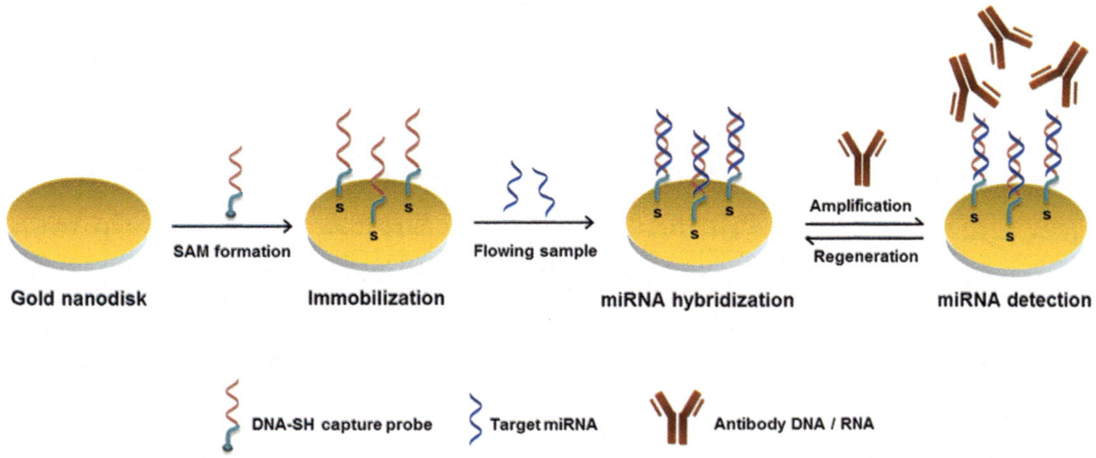

Fig. 10 Complementary hybridization assay and amplification step for miRNA detection

DNA and RNA or single-DNA and -RNA macromolecules. Instead, it interacts with the DNA/RNA duplex and generates a change on the RI that is directly related to the miRNA concentration in the sample.

3.5.5 Immobilization of DNA Capture Probes and Blocking Sensor Surface

1. Prepare a solution of DNA capture probes in 50 mM PBS buffer, and add SH-PEG-NH_2: SH-PEG-COOH (1,1) and 100 nM TCEP. Incubate the mixture for 20 min at 36 °C, 750 rpm, on the thermal mixer.
2. Initiate a continuous flow (18 μL min^{-1}) of DEPC-H_2O.
3. Perform the in situ immobilization of the DNA capture probes using the experimental setup at a constant flow rate.
4. Inject a suitable volume of the DNA capture probe mixture (300 μL) at the same flow rate until stabilization of the sensor signal.
5. Inject a 20 mg mL^{-1} m-PEG-silane solution in ethanol/water (v/v, 95%) at continuous flow rate (9 μL min^{-1}) to *block* the glass surface. LSPR sensor is now functionalized and ready to be used for miRNA detection.

3.5.6 Amplification Step

1. Initiate a continuous flow of SSC 0.5× and wait until the sensor signal is stabilized.
2. Prepare a series of solutions of different concentrations of miRNA-210 in SSC 5× (300 μL, standard solutions).
3. Inject the miRNA-210 standard solutions at 18 μL min^{-1} and wait until the sensor signal stabilizes.
4. Inject the antibody solution (2 μg mL^{-1}) in SSC 0.5×. This step produces a wavelength shift ($\Delta\lambda_{LSPR}$) due to the interaction of the antibody with the DNA/RNA hybrids.
5. Regenerate the sensor surface by injecting 5 mM NaOH solution (18 μL min^{-1}, 60 s) and wait until the signal stabilizes (*see* **Note 3**). At this stage, the sensor surface is ready for the measurement of a second standard solution. Figure 11 shows the sensor response for two samples and the amplification step. The absence of signal with a noncomplementary miRNA (miRNA-205) can be observed (red curve).
6. Plot the sensor response after the amplification step versus miRNA-210 concentration and fit the data to an appropriate curve-fitting model for obtaining the calibration curve (*see* **Note 4**). LOD should be in the desired range for the application. If not, try to optimize immobilization, detection, and/or amplification conditions.

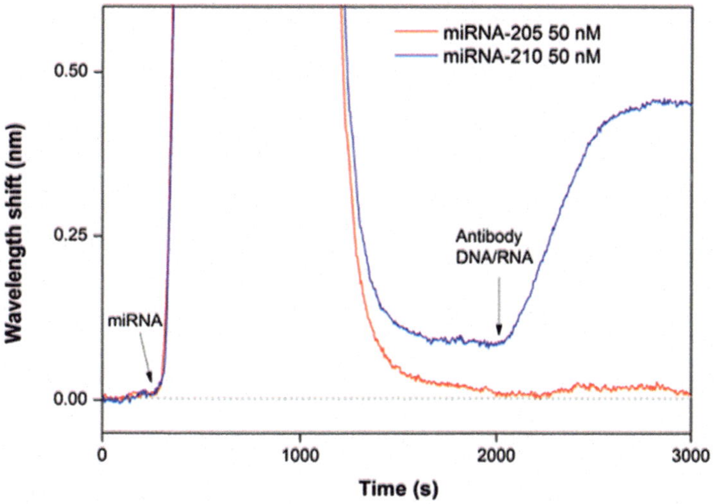

Fig. 11 Real monitoring of miRNA-210 detection performing antibody amplification strategy. The presence of miRNA-210 in solutions and the specific antibody against DNA/RNA hybrids lead to an increase of the sensor response (blue). Absence of changes on the signal is observed when a noncomplementary miRNA (miRNA-205) (red) is used

7. Quantify the concentration of miRNA in samples of unknown concentrations:
   (a) Inject the sample, and wait until the signal stabilizes.
   (b) Inject the antibody solution, wait until the signal stabilizes, and register the value of the sensor response ($\Delta\lambda_{LSPR}$).
   (c) Estimate the concentration of miRNA-210 in unknown samples by evaluating the value of the sensor response ($\Delta\lambda_{LSPR}$) in the calibration curve previously obtained.

3.6 Surface Biofunctionalization for Immunodetection Assays

In the case of colorectal cancer, tumor-associated autoantibodies have become valuable biomarkers for the preclinical diagnosis of cancer. These autoantibodies, which circulate in the blood, are produced by the immune system when the TAA proteins appear. We have selected four specific TAA recombinant proteins present in serum (MAPKAPK3, PIM1, STK4, and GTF2B) for detection through their corresponding autoantibodies (anti-MAPKAPK3, anti-PIM1, anti-STK4, and anti-GTF2B). A schematic representation of the biofunctionalization strategy used to prepare the sensor surface for the antibody detection is shown in Fig. 12.

3.6.1 Cleaning Procedure

Prior to the biofunctionalization process, nanoplasmonic sensor chips must be cleaned following this protocol:

1. Immerse sensor chips with acetone, ethanol, and Milli-Q water, heating for 1 min in a hot plate and sonicating for 1 min in each cycle.

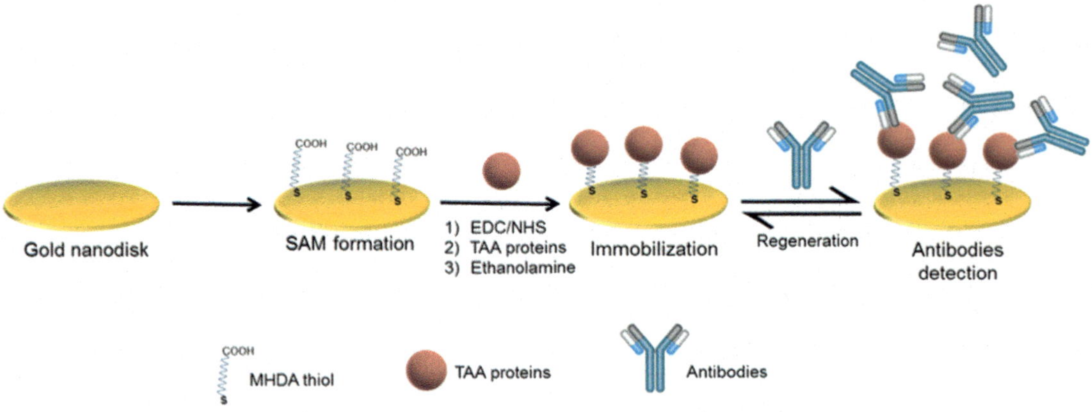

Fig. 12 Schematic representation of the biofunctionalization strategy and the direct immunoassay showing the SAM formation, activation, immobilization, and detection of antibodies

2. Dry with a N_2 stream and place in a UV/Ozone cleaner for 30 min.
3. Finally, rinse sensor chips with ethanol and dry again with N_2 stream.

3.6.2 Immobilization of the Recombinant Proteins

The surface of the sensor chip is prepared for immobilization of the capturing layer formed by the recombinant proteins. This process requires several preparation steps, including the formation of a self-assembled monolayer (SAM), the activation of the SAM, the blocking of the unreacted functional groups, and finally the protein immobilization.

1. Prepare 3 mL of a 50 μM solution of MHDA in ethanol.
2. Immerse the cleaned sensor chips into the solution in order to modify the surface with the SAM and incubate overnight at room temperature (*see* **Note 6**).
3. Rinse sensor chips with ethanol and dry with N_2 stream.
4. Place the nanoplasmonic sensor chips on the experimental setup. Carry out the immobilization of TAA proteins in situ at a continuous flow of Milli-Q water at 20 μL min^{-1}.
5. Inject a mixed solution (200 μL) of EDC/NHS (0.2 M / 0.05 M) in MES buffer for activation of SAM carboxylic groups (*see* **Note 7**).
6. Reduce the flow rate to 10 μL min^{-1} and inject a solution (200 μL) of TAA protein 50 μg mL^{-1} in PBS-1X.
7. Change the flow rate to 30 μL min^{-1} and inject a solution of ethanolamine 1 M, pH 8.5, for 2 min. An example of the real-time sensorgram showing the immobilization procedure is shown in Fig. 13.

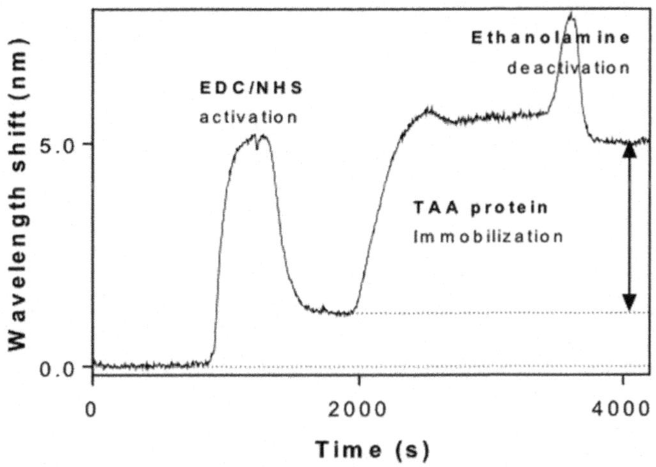

Fig. 13 Real-time LSPR sensorgram showing the in situ immobilization procedure: activation, immobilization of the TAA protein, and blocking of the remaining activated groups

8. After immobilization, settle PBST-0.5 buffer as continuous buffer at 25 μL min^{-1}. The sensor chip is ready to be used for TAA antibody detection.

3.6.3 Direct Immunoassay (Antibody Detection)

After successful biofunctionalization of the sensor surface, different concentrations of TAA antibodies are sequentially flowed onto the sensor surface to obtain a calibration curve of the direct antibody detection. Upon detection of the sensor response to a given concentration, a regeneration process is performed in order to dissociate the protein–antibody interaction, a fact that is evidenced by the recovery of the baseline to the values preceding the antibody injection.

1. Prepare a series of solutions at different concentrations of specific TAA antibodies in PBST-0.5 buffer in a range between 0.5 and 10 μg mL^{-1}. The range is defined by the LOD of each specific immunoassay.
2. Inject 200 μL of each concentration at a flow rate of 30 μL min^{-1} and wait until the sensor signal stabilizes. The bulk effect due to RI changes of each concentration will be observed. A shift in the resonance spectrum ($\Delta\lambda_{LSPR}$) (with respect to the baseline of continuous buffer) is produced by the chemical interaction between the antibodies and the recombinant proteins.
3. Regenerate the sensor surface by injecting 20 mM NaOH at 35 μL min^{-1} for 2 min. The regeneration process allows the total dissociation of the antibody, recovering the same baseline sensor signal. Several cycles of interaction/regeneration must be evaluated for different concentrations of antibodies.

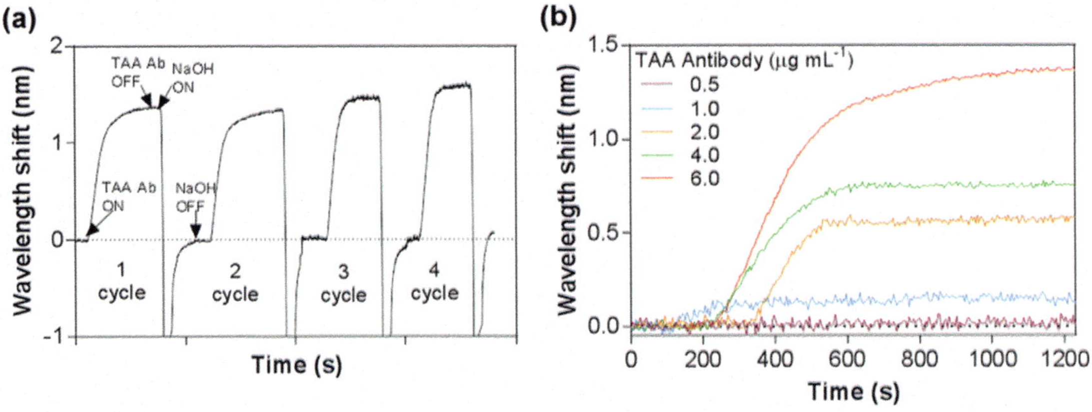

Fig. 14 (**a**) Sensorgram showing wavelength shifts ($\Delta\lambda_{LSPR}$) of successive evaluations of antibody anti-GTF2b 1 μg mL^{-1} and regeneration cycles. (**b**) Real-time sensorgram for different TAA antibody concentrations in PBST-0.5 buffer

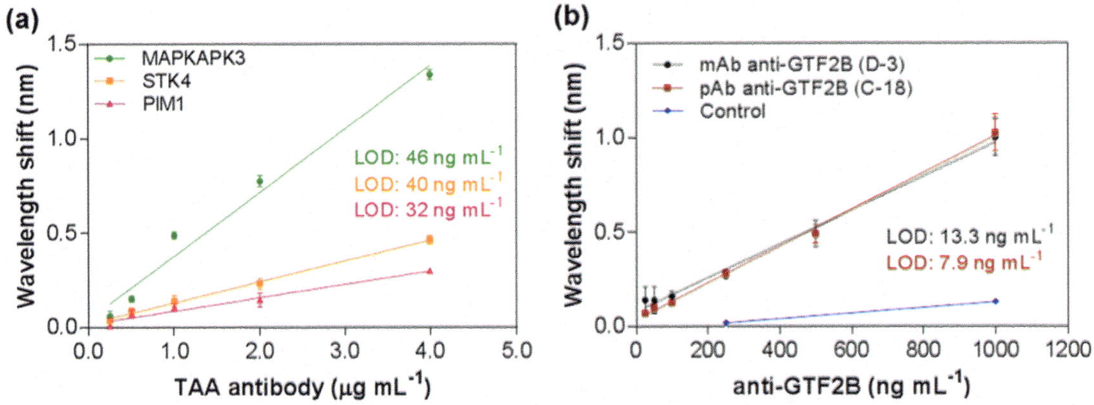

Fig. 15 Calibration curve of TAA antibodies: **(a)** anti-MAPKAPK3 (green), anti-STK4 (orange), and anti-PIM-1 (pink) **(b)** monoclonal mAb anti-GTF2B (D-3) (black) and polyclonal pAb anti-GTF2B (C-18) (red)

Figure 14a shows an example of the wavelength shifts ($\Delta\lambda_{LSPR}$) throughout successive measurement and regeneration cycles.

4. Construct calibration curves for each immunoassay, averaging the concentration values three times.
5. Fit the curve through a linear regression model of sensor response versus TAA antibody concentration (*see* **Note 4**). An example of a real-time sensorgram for different TAA antibody (anti-MAPKAPK3) concentrations in PBST-0.5 is shown in Fig. 14b. Calibration curves of each employed TAA antibody are shown in Fig. 15.
6. Measure the system noise (σ_{SR}), averaging the baseline signal of $\Delta\lambda_{LSPR}$ for 10 min using the corresponding running buffer at a constant flow rate. Calculate the LOD expressed in concentration units (pM–nM), evaluating in triplicate the system noise ($3{\cdot}\sigma_{SR}$) in the calibration curve for each immunoassay.

3.6.4 Detection of TAA Antibodies in Serum

In order to demonstrate the capability of the immunoassays to be transferred to a clinical setting, it is necessary to study the behavior of the biological matrix when flowed across the sensor surface. Because TAA antibodies circulate in blood, we have selected serum to evaluate direct detection immunoassays for each specific TAA antibody. The influence of nonspecific binding onto the surface must be assessed, owing to the potential interactions with other components of the serum matrix that can affect the sensor signal analysis. Therefore, a *blocking* of the sensor surface using *antifouling* agents is a mandatory step to avoid or reduce these interferences (*see* **Note 5**).

1. Inject 200 μL of an undiluted serum sample, in the absence of analyte, and verify there is no significant sensor response ($\Delta\lambda_{LSPR}$ ~0 nm). The presence of a sensor response means that nonspecific sensor binding is taking place; therefore a dilution of the serum sample is required.
2. Prepare and inject solutions of diluted serum (e.g., 5%, 10%, or 20%), using the same buffer employed for the calibration curve of the direct immunoassays. If all the dilutions produce a nonspecific response, proceed to add the blocking agent using the same dilutions previously tested.
3. Inject a solution of PLL-g-PEG (0.5 mg mL^{-1}) in HEPES buffer, pH 7.5, to coat the sensor surface in order to reduce nonspecific adsorptions of serum.
4. Compare the same serum dilutions of (2.) and select the least diluted sample that produces the smallest sensor response. Notice in Fig. 16a that in all cases, serum dilutions after the addition of blocking agent show significant reduction of the sensor signal associated with nonspecific adsorption. In our studies, the most effective blocking was achieved for the 10% diluted serum, with a decrease of 83.36% of the nonspecific sensor response.
5. In turns inject the following solutions into the sensor surface: (i) serum dilution 10%, (ii) serum dilution 10% with a known concentration of a specific TAA antibody, and (iii) buffer containing known concentration of a specific TAA antibody.
6. Assess the effectiveness of the blocking step (i.e., PLL-g-PEG) by comparing the sensor signal response of dilutions *ii* and *iii* of (5.) to the sensor response prior to the blocking step. Figure 16b shows no response for the 10% serum after the blocking step (orange), and all signals containing the antibody (blue, pink, green) showed the same signal response. This means that TAA antibody detection assay was not affected by the added blocking step.

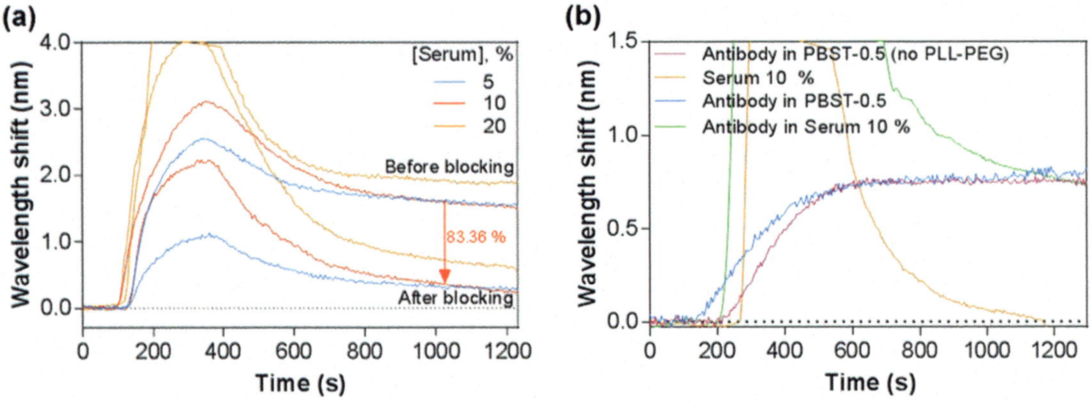

Fig. 16 (**a**) Sensorgram showing the effect of the blocking step using PLL-g-PEG (0.5 mg mL^{-1}) at different concentrations of diluted (5% (blue), 10% (red), and 20% (orange)). (**b**) Real-time sensorgram showing the effect of the PLL-g-PEG blocking step in the detection of TAA antibodies: pink and blue lines show the detection of anti-MAPKAPK3 (4 μg mL^{-1}) without and with PLL-g-PEG layer, respectively; orange line shows serum diluted 10% without analyte; green line shows the detection of anti-MAPKAPK3 (4 μg mL^{-1}) in serum diluted 10%

4 Notes

1. All the buffers and other solutions for miRNA detection were prepared using DEPC-H_2O (Milli-Q water incubated in continuous agitation overnight with 0.1% DEPC and autoclaved at 121 °C for 1 h) in order to inactive RNase enzyme in water. All solid plastic and glass materials were autoclaved at 121 °C for 1 h. Moreover, the microfluidic system must be cleaned following RNase-free protocol.
2. Oxygen plasma treatment can be replaced by UV/ozone plasma treatment for 30 min. Both oxidative pretreatments produce hydrophilic surfaces.
3. Regeneration strategy depends on the specific interaction RNA/DNA, or protein-antibody. The ideal regeneration solution should disrupt this interaction, preserving the bioreceptors, without deactivating or denaturing it. However, all regeneration solutions cause some degree of damage on the bioreceptor layer, limiting the number of cycles that the sensor chip can be reused. Usually, low pH (0.01–1 M HCl, 100 mM phosphoric acid), high salt concentrations (3 M $MgCl_2$), high pH (10–100 mM NaOH), or chaotropic solutions, alone or combined with surfactant reagents, are employed [23].
4. It is critical to choose an appropriate curve-fitting model to minimize quantitation error of concentration values. The fitting model used for the target miRNA is *one-site-specific binding* $y = \frac{A \cdot X}{(B+X)}$, GraphPad (GraphPad Software Inc., USA),

where y is the sensor response, X is the concentration of the target analyte, A is the extrapolated maximum number of bioreceptors in the surface, and B is the equilibrium binding constant, which corresponds to the analyte concentration needed to achieve half-maximum bioreceptors occupied at equilibrium.

For immunodetection assays of TAA antibodies, lineal regression fitting of the data $y = mx + b$ is performed using OriginPro (OriginLab Corp., USA) where y is the sensor response, X is the concentration of the target analyte, and m defines the sensitivity of the system.

5. In label-free optical biosensors, it is extremely important to avoid nonspecific binding in order to prevent false-positive results from the matrix components and to generate a more biocompatible and antifouling biolayer. Thiolated PEGs (SH-PEGs) carrying different functional groups (NH_2, COOH) act as surface-blocking agents, avoiding nonspecific adsorptions on the gold sensor surface, and as horizontal spacer, regulating DNA capture probe density. Different ratios of SH-PEGs:SH-DNA probes could be tested to obtain the optimal probe density with antifouling properties. PEG-silanes can be used to modify the glass surface and suppress the non-specific binding of charged molecules [24, 25].
6. Generally, SAMs are made by immersing a sensor chip into a dilute solution of alkanethiol in ethanol and are often left over 12–72 h at room temperature in the absence of light to avoid problems due to oxidation or photoinduced processes, which can affect terminal groups and lead to disorder and multilayer formation [26].
7. EDC is hygroscopic and rapidly can oxidize in air. EDC/NHS solution must be prepared immediately before to use.

References

1. Torre LA, Siegel RL, Ward EM et al (2016) Global cancer incidence and mortality rates and trends—an update. Cancer Epidemiol Biomarkers Prev 25(1):16–27
2. Oltean S, Bates D (2014) Hallmarks of alternative splicing in cancer. Oncogene 33:5311
3. Jansson MD, Lund AH (2012) MicroRNA and cancer. Mol Oncol 6:590–610
4. Esteller M (2005) Aberrant DNA methylation as a cancer-inducing mechanism. Annu Rev Pharmacol Toxicol 45:629–656
5. Hrustincova A, Votavova H, Merkerova MD (2015) Circulating microRNAs: methodological aspects in detection of these biomarkers. Folia Biol 61:203
6. Lu J, Getz G, Miska EA et al (2005) MicroRNA expression profiles classify human cancers. Nature 435:834
7. Zhang B, Pan X, Cobb GP et al (2007) microRNAs as oncogenes and tumor suppressors. Dev Biol 302:1–12
8. He R-Q, Cen W-L, Cen J-M et al (2018) Clinical Significance of miR-210 and its prospective signaling pathways in non-small cell lung cancer: evidence from gene expression omnibus and the cancer genome atlas data mining with 2763 samples and validation via real-time quantitative PCR. Cell Physiol Biochem 46:925–952

9. Li J-H, Sun S-S, Li N et al (2017) MiR-205 as a promising biomarker in the diagnosis and prognosis of lung cancer. Oncotarget 8:91938–91949
10. Soler M, Estevez M-C, Villar-Vazquez R et al (2016) Label-free nanoplasmonic sensing of tumor-associate autoantibodies for early diagnosis of colorectal cancer. Anal Chim Acta 930:31–38
11. Reuschenbach M, Dörre J, Waterboer T et al (2014) A multiplex method for the detection of serum antibodies against in silico-predicted tumor antigens. Cancer Immunol Immunother 63:1251–1259
12. Barderas R, Villar-Vázquez R, Fernández-Aceñero MJ et al (2013) Sporadic colon cancer murine models demonstrate the value of autoantibody detection for preclinical cancer diagnosis. Sci Rep 3:2938
13. Haes AJ, Zou S, Schatz GC et al (2004) A nanoscale optical biosensor: the long range distance dependence of the localized surface plasmon resonance of noble metal nanoparticles. J Phys Chem B 108:109–116
14. Jung LS, Campbell CT, Chinowsky TM et al (1998) Quantitative interpretation of the response of surface plasmon resonance sensors to adsorbed films. Langmuir 14:5636–5648
15. Babel I, Barderas R, Díaz-Uriarte R et al (2009) Identification of tumor-associated autoantigens for the diagnosis of colorectal cancer in serum using high density protein microarrays. Mol Cell Proteomics 8:2382–2395
16. Babel I, Barderas R, Diaz-Uriarte R et al (2011) Identification of MST1/STK4 and SULF1 proteins as autoantibody targets for the diagnosis of colorectal cancer by using phage microarrays. Mol Cell Proteomics 10 (3):M110.001784. https://doi.org/10.1074/mcp.M110.001784
17. Estevez M-C, Otte MA, Sepulveda B et al (2014) Trends and challenges of refractometric nanoplasmonic biosensors: a review. Anal Chim Acta 806:55–73
18. Otte MA, Estevez MC, Regatos D et al (2011) Guiding light in monolayers of sparse and random plasmonic meta-atoms. ACS Nano 5:9179–9186
19. Johnson PB, Christy R-W (1972) Optical constants of the noble metals. Phys Rev B 6:4370
20. Otte MA, Estévez MC, Carrascosa LG et al (2011) Improved Biosensing Capability with Novel Suspended Nanodisks. J Phys Chem C 115:5344–5351
21. Fredriksson H, Alaverdyan Y, Dmitriev A et al (2007) Hole–mask colloidal lithography. Adv Mater 19:4297–4302
22. Sakao Y, Nakamura F, Ueno N et al (2005) Hybridization of oligonucleotide by using DNA self-assembled monolayer. Colloids Surf B: Biointerfaces 40:149–152
23. Schasfoort RB (2017) Handbook of surface plasmon resonance. In: Royal Society of Chemistry, 2nd ed., Croydon, United Kingdom
24. Krishnan S, Weinman CJ, Ober CK (2008) Advances in polymers for anti-biofouling surfaces. J Mater Chem 18:3405–3413
25. Jo S, Park K (2000) Surface modification using silanated poly (ethylene glycol) s. Biomaterials 21:605–616
26. Love JC, Estroff LA, Kriebel JK et al (2005) Self-assembled monolayers of thiolates on metals as a form of nanotechnology. Chem Rev 105:1103–1170

Chapter 11

A Microfluidic E-Tongue System Using Layer-by-Layer Films Deposited onto Interdigitated Electrodes Inside a Polydimethylsiloxane Microchannel

Maria L. Braunger, Cristiane M. Daikuzono, and Antonio Riul Jr

Abstract

An electronic tongue (e-tongue) is a multisensory system employed in the analysis of liquid samples, transforming the raw data into specific recognition patterns through computational and statistical analysis. Distinct types of e-tongues have been reported in the literature, with a plethora of applications in several areas of research. Recently, e-tongues have been integrated into microfluidic devices, which offer advantages such as the use of continuous flow for faster and more accurate analysis, and reduction in size of the devices and volumes for sampling and discharge, which in turn reduces waste and cost. Here we describe the procedures and methodologies recently used in our research group in the development of a microfluidic e-tongue sensing system.

Key words Electronic tongue, Microfluidic device, Liquid analysis, Thin film, Layer-by-layer

1 Introduction

Microfluidic e-tongues [1] have been used in several analyses to distinguish basic flavors, coffee brands, presence of gluten in foodstuff, and soil analysis [2–4]. Impedance spectroscopy was used in data acquisition as it is a fast, noninvasive method that also avoids polarization effects in samples, with impedance data analyzed through computational or statistical methods to help visualize and distinguish sample differences. 1 mM Aqueous solutions of various flavors including HCl (sour), NaCl (salt), caffeine (bitter), sucrose (sweet), and L-glutamic acid monosodium salt hydrate (umami) were analyzed by impedance spectroscopy at 1 kHz, with good distinction of all flavors [1].

The first microfluidic e-tongue using impedance spectroscopy [1] paved the way to distinct new types of microfluidic applications, with interesting developments. Alessio et al. [2] used a microfluidic e-tongue to distinguish different coffee brands (traditional,

Jessica E. Fitzgerald and Hicham Fenniri (eds.), *Biomimetic Sensing: Methods and Protocols*, Methods in Molecular Biology, vol. 2027, https://doi.org/10.1007/978-1-4939-9616-2_11,

gourmet, organic, and premium) demonstrating the device ability to separate selected gourmet, organic, and premium samples from unselected grains. Another study by Daikuzono et al. [3] applied a microfluidic e-tongue for the detection of different concentrations of gliadin in ethanol solution to distinguish gluten-free food stuffs from those containing gluten. Through computational analysis (Interactive Document Map, IDMAP), it was possible to observe different concentrations of gliadin and the successful distinction of gluten-free foodstuff from those containing gluten. Furthermore, microfluidic e-tongues can be used in soil analysis to monitor the presence of macronutrients and ensure a rapid soil examination to assist precision agriculture [4]. Shimizu et al. developed a novel microfluidic e-tongue consisting of a single piece of polydimethylsiloxane (PDMS) with five stainless steel microwires, which were short-circuited and coated with distinct metal oxides. The innovation of this approach was the effortless replacement of the sensing units in case of damage or contamination [5].

Various methods can be used to assemble thin films as sensing units for e-tongue applications. Therefore, the layer-by-layer (LbL) technique is a versatile, non-expensive, and simple way for multilayer formation on solid substrates, either on surfaces or inside microchannels [6]. The multilayer deposition of charged inorganic colloids by consecutive adsorption was initially proposed by Iler [7]. Based on Iler's method, Decher highlighted that the LbL assembly by physical adsorption from aqueous solutions is an even more general approach for the fabrication of multicomponent films on solid supports, where materials can be selected from small organic molecules, polymers, natural proteins, inorganic particles, clays, and colloids [8]. The LbL technique consists of immersing a solid substrate in an aqueous solution containing the material to be adsorbed. Subsequently, the substrate is rinsed to remove material that is loosely bound, allowed to dry, and then immersed in another polyelectrolyte solution having opposite charge to the material initially adsorbed. Then, multilayered thin films are formed by alternating adsorbed cationic and anionic molecular bilayers. The wide use of the LbL technique in numerous applications has led to the development of new LbL techniques that can be organized in five distinct categories [6]: immersive, spin, spray, electromagnetic, and fluidic assembly.

Herein, we describe the procedures and methodologies recently used in our research group to develop a microfluidic e-tongue system. Briefly, it comprises an array of four sensing units (Fig. 1a) using LbL films deposited onto gold interdigitated electrodes (IDE) inside a PDMS microchannel (Fig. 1b). Authors fabricated the IDEs at the Brazilian Nanotechnology National Laboratory (LNNano/CNPEM) and PDMS is a polymer widely used in micro- and nanofabrication of devices due to its good optical transparency, gas permeability, chemical inertness,

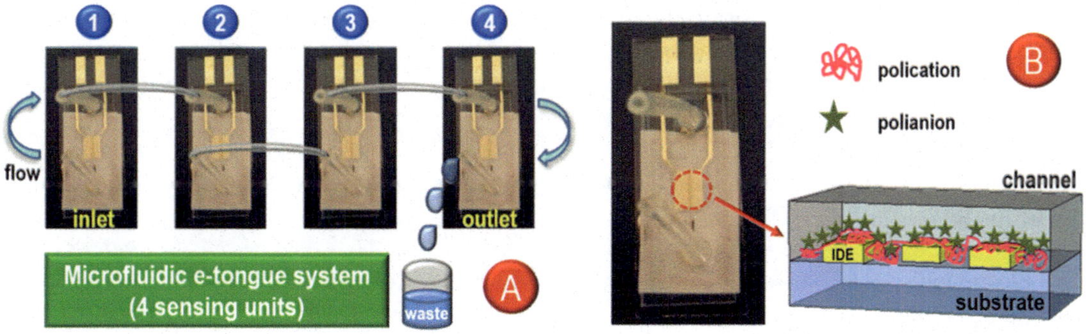

Fig. 1 (**a**) Schematic representation of a microfluidic e-tongue system comprising four sensing units to measure a specific analyte. (**b**) LbL films deposited onto gold IDEs inside a PDMS microchannel

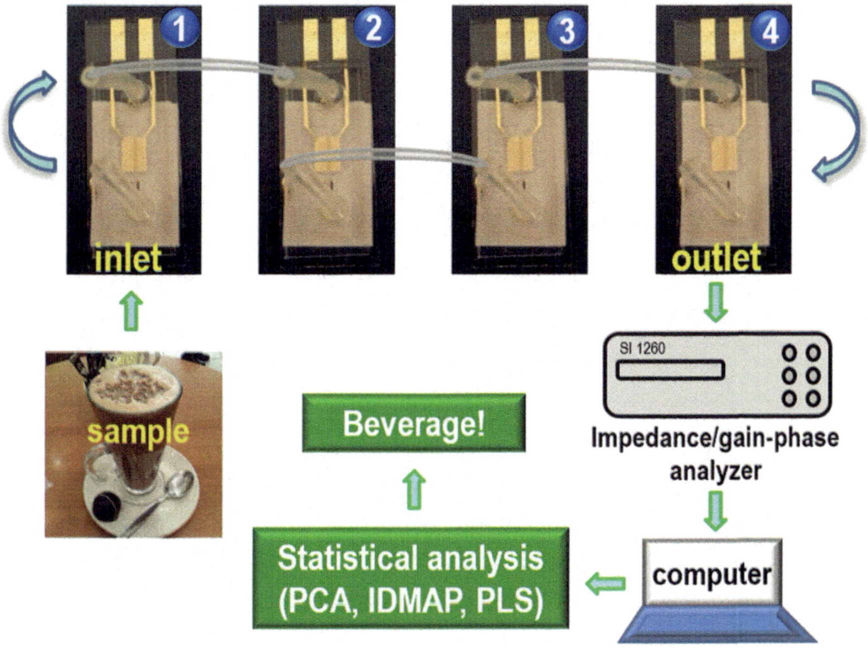

Fig. 2 Sequence of operation from electrical impedance measurements of a sample to the statistical analysis

biocompatibility, and reproducibility. In addition, the PDMS surface can be easily modified with oxygen plasma, allowing the rapid sealing of the device on glass slides.

The raw impedance data from the e-tongue system can be further analyzed with principal component analysis (PCA) (Fig. 2). PCA is a statistical method widely used to reduce the dimensionality of the original data without losing information. This method consists of rewriting the coordinates of the data set (input matrix) into a new orthogonal axis system called principal components, avoiding redundancy of information. From the linear combination of the original variables the data can be represented by a smaller number of descriptive factors, thus facilitating the visualization of the samples.

2 Materials

2.1 Electrodes and Microchannel Fabrication

1. Gold interdigitated electrodes (IDE) deposited onto glass slides (BK7), using chromium or titanium as adhesion layer: comprised by 30 pairs of digits, 3 mm length, and having 40 μm in both width and inter-electrode gap.
2. Microchannels fabricated using PDMS: 490 μm wide, 50 μm high, and 12.5 mm length, sealed onto IDE using oxygen plasma.
3. Ultrasonicator.
4. Deionized (DI) water.
5. Extran soap.
6. Piranha solution.
7. Spin coater.
8. Hexamethyldisilazane (HMDS).
9. Clariant AZ 4210 photoresist.
10. Hot plate.
11. Acetate photomask.
12. UV lamp: 300 W.
13. Photoaligner.
14. Clariant K400 developer.
15. Oxygen plasma.
16. Pressure chamber: 100 mTorr chamber at 100 W.
17. Balzers BA510 sputtering equipment.
18. Chromium or titanium layers: 30 nm thick.
19. Gold layer: 120 nm.
20. Acetone.
21. SU-8 photoresist.
22. SU-8 developer.
23. Isopropyl alcohol.
24. Sylgard Kit for PDMS: Base and curing agent (10:1).
25. Silicon dioxide (SiO_2).
26. Chemical vapor deposition instrument.
27. Potassium hydroxide.
28. Methanol.

2.2 Thin Films (See Note 1)

2.2.1 Cationic Layer

1. Polyallylamine hydrochloride (PAH).
2. Poly(diallyldimethylammonium chloride) (PDDA).
3. Polyethylenimine (PEI).

2.2.2 Anionic Layer

1. Poly(3,4-ethylenedioxythiophene):poly(styrenesulfonate) (PEDOT:PSS).
2. Polypyrrole (PPy).
3. Nickel(II) phthalocyanine-tetrasulfonic acid tetrasodium salt (NiTsPc).
4. Copper phthalocyanine-3,4′,4″,4‴-tetrasulfonic acid tetrasodium salt (CuTsPc).
5. Montmorillonite clay (MMt).

2.3 Analyte Samples

1. The analyte choice depends on the desired application. Different kinds of samples have been reported in the literature so far, such as water, coffee, wine, pharmaceuticals, and soil (*see* **Note 2**).

3 Methods

Photolithography was the method of choice to fabricate the gold IDEs onto glass substrates and the microchannel mold. The procedure consists of UV radiation exposure to transfer a predetermined geometric pattern of a mask to a photosensitive chemical compound (photoresist), which was previously deposited on a glass slide. After exposure, the substrate is immersed in a developer, which is used to remove areas that have not been polymerized (negative photoresist) or remove regions that have been decomposed (positive photoresist).

3.1 Interdigitated Electrode Fabrication

1. Clean the glass substrates using an ultrasonic bath at 50 °C with deionized water and Extran soap for 5 min.
2. Submerge substrates in a piranha solution, and then submerge in a bath with DI water and subsequent heating at 120 °C for some minutes to allow them to dry.
3. Apply HMDS using a spinner at 100.8 × *g* for 30 s.
4. Apply Clariant AZ 4210 photoresist using a spinner at 100.8 × *g* for 30 s.
5. Heat the substrate on a hot plate at 95 °C for 15 min to speed up solvent evaporation.
6. Apply the acetate photomask over the substrate assembly under 300 W UV radiation for 30 s (*see* **Note 3**).

7. The development of the photoresist is made by dipping the substrate in Clariant K400 developer for 30 s.
8. After the development, expose the substrates to oxygen plasma to remove possible residues and assure good adhesion of the metal layer. The exposure to plasma takes 3 min and is carried on in a 100 mTorr chamber and 100 W power.
9. The metallic layers are deposited by sputtering. The metallic film is composed of a 30 nm chromium (Cr) or titanium (Ti) layer to facilitate the adhesion of other metals, and a second 120 nm layer of gold (Au).
10. The last step is called liftoff, when the glass substrates are submerged in acetone for 1 h to remove the photoresist that served as the mold for the metallic electrodes. The metal film that is attached onto the photoresist is also removed in this step, leaving only the metal film that has been in direct contact to the glass.

3.2 Microchannel Mold Fabrication

Similar to the IDE fabrication method, photolithography was also applied to create the mold for microchannel fabrication. Some specificities are outlined here:

1. Spin-coat the negative photoresist SU-8 on the substrate at $111.2 \times g$ for 30 s to obtain ~50 μm photoresist thickness.
2. Heat the substrate on a hot plate at 65 °C for 5 min and at 95 °C for 15 min.
3. Expose the substrate to UV radiation for 50 s.
4. Bake the substrate at 65 °C for 5 min and at 95 °C for 15 min.
5. Submerge the substrate in SU-8 developer for development.
6. Clean the substrate with isopropyl alcohol to remove excess at the edges, and place once more in SU-8 developer for 40 s.

3.3 PDMS Microchannel Fabrication

1. Affix the microchannel mold to a metal support along with silicone hoses (internal diameter equal to 1 mm) (*see* **Note 4**).
2. Once the microchannel mold is ready, prepare PDMS using a Sylgard kit, consisting of a base and a curing agent (10:1).
3. Keep the blend of this commercial elastomer under vacuum pressure for ~1 h until air bubbles disappear.
4. Carefully pour the viscous liquid into the mold. After PDMS is placed in the SU-8 template, leave it on a hot plate at 100 °C for 1 h. During the curing process, the PDMS becomes an elastomer, taking the shape of the mold and adhering to the silicone hoses.
5. After cooling, peel the PDMS devices from the mold with the aid of a scalpel.

3.4 Sealing the Devices

1. Before sealing the substrate with the PDMS microchannel, deposit 50 nm of SiO_2 onto IDEs by chemical vapor deposition to improve the sealing process.
2. Rinse the IDEs with isopropanol.
3. Immerse PDMS channels in a solution of deionized water, potassium hydroxide, and isopropanol (1:1:8 v/v/v), and then rinse with methanol.
4. Expose IDEs and PDMS microchannels to oxygen plasma (100 mTorr and 70 W) for 15 s.
5. After plasma treatment, immediately place the PDMS microchannel in contact with the IDE substrate and manually press for a few seconds, sealing irreversibly the device.

3.5 Layer-by-Layer Deposition

To build LbL films inside microchannels we use the fluidic assembly, in which it is also possible to choose between dynamic (fluid flow) [9] and static incubation (solutions remain in static contact with the substrate walls inside the microchannel) [6]. We usually apply the dynamic LbL deposition as follows:

1. Prepare all polyelectrolytes using deionized water.
2. Alternate aqueous solutions of the materials from a specified time period for each solution inside the microchannels at 1000 $\mu L.h^{-1}$ flow (*see* **Notes 5** and **6**).
3. Remove the aqueous solution from the channel with the aid of a vacuum pump between each deposited layer.
4. Let the adsorbed layer dry for 5 min and repeat the process for the next cationic or anionic layer.
5. After thin-film deposition, rinse the devices before starting data acquisition in order to avoid possible interference due to removal of excess material loosely bound on the electrodes.

3.6 Impedance Spectroscopy Measurements

1. Prepare all analyte solutions using deionized or ultrapure water.
2. Make proper connection between electrodes and impedance analyzer using Bayonet Neill-Concelman (BNC) connectors (*see* **Note** 7).
3. These measurements can be performed with the liquid samples injected inside the microchannels in static mode or at different flow rates, depending on the application and the sample behavior (*see* **Notes 8** and **9**). Use a syringe pump to obtain the desired flow. It is usually from 1000 to 5000 $\mu L.h^{-1}$ for regular applications in our research group.
4. Electrical measurements are usually made using 20 to 25 mV of amplitude in the frequency range 1–10^6 Hz using a Solartron 1260A with an impedance/gain-phase analyzer coupled to a 1296A dielectric interface (*see* **Note 10**).

5. Before measuring the analyte, perform electrical measurements using distilled or deionized water to obtain a standard capacitance plot for each sensing unit, and also check the stability of the electrical measurements.
6. For each analyte solution, take five scans from 1 to 10^6 Hz and use the last one as the final data (*see* **Note 11**).
7. After measuring the analyte, remove the liquid from the microchannel using a syringe or a vacuum pump. Pass distilled or deionized water to clean the interior of the microchannel.
8. Perform an electrical measurement using distilled or deionized water to evaluate if the result corresponds to the original one made before measuring the analyte (*see* **Note 12**).
9. Repeat the process at least three times for each analyte sample in order to obtain independent measurements for the PCA analysis.

3.7 Principal Component Analysis

For electronic tongue applications the real part of the capacitance, C', was usually considered, and principal component analysis (PCA) graphs are obtained from C' at a fixed frequency (~kHz) for each sensing unit.

4 Notes

From several trials over the past 15 years, we have compiled some tips for the student or researcher who is interested in using similar procedures to those presented herein.

1. The materials forming the LbL films should be chosen for the manufacture of at least four different sensing units, thus forming the e-tongue system. Another important recommendation is to use materials having distinct electrical characteristics, thus facilitating the differentiation of the samples.
2. Some analytes may exhibit low water solubility, which may result in material decantation in the measured solution. A reduced concentration in the best scenario will mask results, making it difficult to obtain a correct fingerprint of the measured analytes. Another possible issue in this case is the contamination of the sensing units, decreasing the lifetime of the e-tongue system.
3. During the step involving UV exposure through the photomask, the positive photoresist will be exposed to light and the pattern of the photomask will be transferred to it. A photoaligner was used to best fit the photomask to the substrate.
4. Regarding the PDMS microchannel fabrication, the hoses are fixed using acrylic supports and they aligned with the inlet and

outlet reservoirs, connecting the system to the external environment. It is necessary to fit a nail into one end of the hose before aligning with the mold, thus preventing the polymerization of PDMS inside the hose. After the PDMS polymerization, just remove the nail with the aid of tweezers.

5. Some caution must be taken in the choice of materials and during the deposition process of the thin films onto the IDEs: check if the thin-film adhesion was successful and always be aware of possible removal of the electrodes during the deposition process.
6. The choice of materials is also important according to the type of analyte studied. For example, a material may present low sensitivity response to a certain electrolyte; nonetheless its use may be important for the analysis of other analytes and also in the fingerprint formation of the sample (recognition pattern).
7. It is not required to place the electrodes inside a metallic box, but it is encouraged to avoid electric noise during data acquisition (it will improve your results).
8. In our recent paper regarding soil analysis [4], for example, the choice of flow instead of static measurements was made to avoid a possible soil decantation inside the microchannel, which could lead to contamination of the sensing units, thus compromising the reusability of the device.
9. During the measurements always check for possible leaks or clogging in the microchannel devices, and the possible removal of the metal part of the IDEs due to fluid flow.
10. The low voltage value applied for the impedance measurements is chosen to avoid the electrolysis of water, which could modify the data or hinder the observed results. Analyzing the impedance spectrum at low frequencies (<100 Hz) will have a strong influence from the electrical double layer. The frequency at kHz region is related with the electrode/electrolyte interface and at higher frequencies ($>10^4$ Hz) the impedance data is related with the electrode geometry.
11. Depending on the sensing unit/analyte pair impedance scans require distinct times to stabilize the system due to double-layer formation. Therefore, it is highly recommended to check the time needed for a stable impedance signal before starting the data acquisition.
12. Cross-contamination of the sensing units is also possible; therefore, always rinse the devices with deionized or distilled water between distinct measurements to avoid that issue.

Acknowledgments

Authors are grateful for the financial support of the Brazilian research funding agencies: FAPESP (Grants n° 2014/03691-7, 2015/14836-9 and 2017/11277-4), CAPES, and CNPq. They also thank LNNano/CNPEM (LMF project n° 21340).

References

1. Daikuzono CM, Dantas CAR, Volpati D et al (2015) Microfluidic electronic tongue. Sensors Actuators B Chem 207:1129–1135
2. Alessio P, Constantino CJL, Daikuzono CM et al (2016) Analysis of coffees using electronic tongues. In: Mendez MR (ed) Electronic noses and tongues in food science. Academic Press, pp 171–177
3. Daikuzono CM, Shimizu FM, Manzoli A et al (2017) Information visualization and feature selection methods applied to detect gliadin in gluten-containing foodstuff with a microfluidic electronic tongue. ACS Appl Mater Interfaces 9:19646–19652
4. Braunger ML, Shimizu FM, Jimenez MJM et al (2017) Microfluidic electronic tongue applied to soil analysis. Chemosensors 5:1–10
5. Shimizu FM, Todão FR, Gobbi AL et al (2017) Functionalization-free microfluidic electronic tongue based on a single response. ACS Sensors 2:1027–1034
6. Richardson JJ, Björnmalm M, Caruso F (2015) Technology-driven layer-by-layer assembly of nanofilms. Science 80(348):2491. https://doi.org/10.1126/science.aaa2491
7. Iler RK (1966) Multilayers of colloidal particles. J Colloid Interface Sci 21:569–594
8. Decher G (1997) Fuzzy nanoassemblies: toward layered polymeric multicomposites. Science 277:1232–1237
9. Kim H-J, Lee K, Kumar S, Kim J (2005) Dynamic sequential layer-by-layer deposition method for fast and region-selective multilayer thin film fabrication. Langmuir 21:8532–8538

Chapter 12

Molecularly Imprinted Polymer Thin-Film Electrochemical Sensors

Vera L. V. Granado, M. Teresa S. R. Gomes, and Alisa Rudnitskaya

Abstract

Preparation of potentiometric and amperometric sensors with thin-film membranes based on molecularly imprinted polymers (MIP) is described. Spherical MIP microparticles with diameter below 1 μm are suitable for incorporation into the sensing membrane by the deposition of a conducting polymer on the electrode surface. This is achieved through electropolymerization from the suspension of MIP particles in monomer solution. Procedures of the synthesis of MIP particles, preparation of sensing membranes, and analytical application of potentiometric and amperometric sensors with MIP-modified membranes are described.

Key words Molecularly imprinted polymers, Precipitation polymerization, Potentiometric sensors, Amperometric sensors, Electropolymerization, Conducting polymers, Plasticized polyvinyl chloride

1 Introduction

Molecular imprinting has gained popularity as a technique of synthesizing polymer materials with chemically selective recognition sites [1, 2]. Molecular imprinting is achieved via the polymerization of the monomer mixture in the presence of a target molecule or template in the inert solvent. After polymerization, the template is removed from the polymer matrix, thus leaving cavities or specific binding sites in the resulting material, creating the molecularly imprinted polymers (MIPs). The high stability of MIPs, as well as their versatility, i.e., the possibility to synthesize polymers for practically any analyte, including low- and high-molecular-weight compounds and microorganisms, renders them attractive artificial ligands or receptors for chemical sensing. One of the challenging issues of this sensing technique is the integration of the MIP with the sensor transducer. Several approaches can be employed, including in situ polymerization by grafting [3–5], preparation of imprinted conducting polymers or sol-gels on the sensor surface [6–8], and deposition of the pre-polymerization mixture [9] or of a

Jessica E. Fitzgerald and Hicham Fenniri (eds.), *Biomimetic Sensing: Methods and Protocols*, Methods in Molecular Biology, vol. 2027, https://doi.org/10.1007/978-1-4939-9616-2_12, © Springer Science+Business Media, LLC, part of Springer Nature 2019

composite that is incorporated with the previously synthesized MIP onto the sensor surface [10, 11].

Synthesis of MIP directly on the sensor surface has been associated with low sensitivity or selectivity of the resulting sensing layer [12]. Therefore, the method described here consists of the electrosynthesis of a conducting polymer from the suspension of the MIP in monomer solution. This results in the entrapment of MIP particles in the conducing polymer matrix on the sensor surface upon polymerization [13]. An appealing aspect of this approach is the possibility to decouple MIP synthesis and immobilization, thus enabling better optimization of each individual step. Synthesis of MIP by precipitation polymerization produces polymer particles with sizes ranging from hundreds of nanometers to several microns. Precipitation polymerization has the advantage of producing regular-shaped polymer particles with well-defined binding sites on the surface [14, 15] that can be directly used for the preparation of the sensing membranes.

2 Materials

Prepare all solutions using ultrapure water (prepared by purifying deionized water, to attain a conductivity of 18 MΩ-cm at 25 °C) and analytical grade reagents. Prepare all solutions before use and store all reagents according to the indication of the manufacturer. Follow all waste disposal regulations when disposing of waste materials. Indicated amounts of reagents are for synthesizing approximately 0.5 g of MIP.

2.1 Polymerization

1. Pre-polymerization mixture: Dissolve 1.5 mmol of the monomer, methacrylic acid (MAA), and template in 40 mL of acetonitrile. Template concentration may vary, but it is recommended to be between 2 and 7.5 times lower than that of the monomer. Pre-polymerization mixture should be prepared before polymerization and used immediately.
2. Cross-linker: Trimethylolpropane trimethacrylate (TMPTMA).
3. Radical initiator: 2,2′-Azobis(2-methylpropionitrile) (AIBN).
4. Water bath with temperature control.
5. 50 mL jacketed glass reactor connected to the water bath through two hose barbs, and with two side arms used for connecting reactor to nitrogen line and for introduction of reagents.
6. Coil condenser cooled by water to avoid solvent evaporation.
7. Magnetic stirrer and stir bar. An assembly is depicted in Fig. 1.

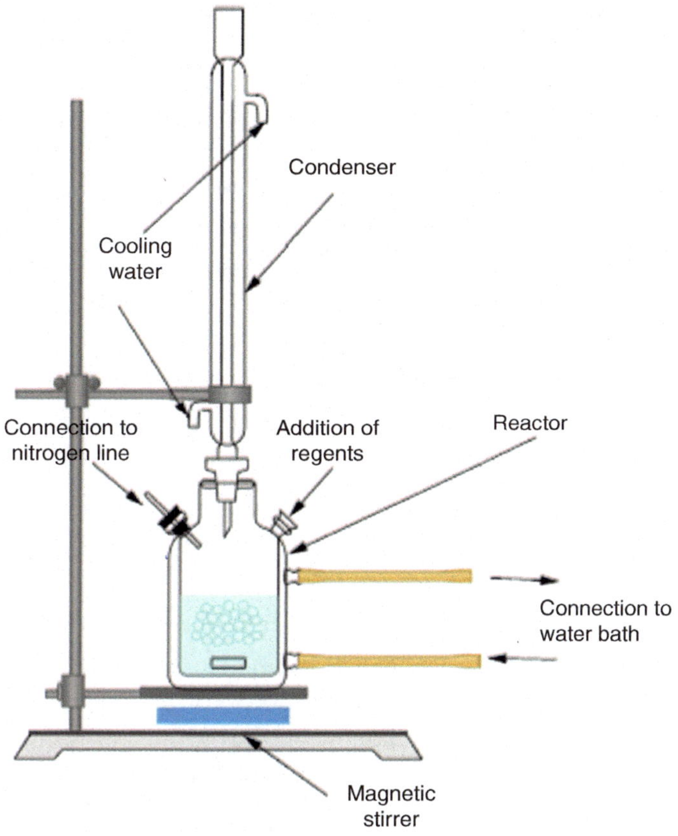

Fig. 1 Schematic setup for the synthesis of MIP microparticles

2.2 Template Removal from MIP Particles

1. Acetonitrile.
2. Laboratory centrifuge with at least 13,000 × *g*.
3. 10 × 50 mm cellulose extraction thimble.
4. Soxhlet extractor setup with main chamber to fit cellulose extraction thimble and reflux condenser cooled by water and heating element.
5. Methanol with 10% v/v of acetic acid, 60–80 mL.
6. Methanol for particle washing.
7. Laboratory oven.

2.3 Electrode Preparation

1. Noble metal (gold or platinum) working electrode, i.e., disk electrode, thin-film electrode manufactured by photolithography, or thick-film electrode manufactured by screen printing.
2. Water and ethanol for washing.
3. 4000-grit abrasive paper and 6 μm diamond polish for electrode polishing.

4. Solutions: 6 M sulfuric acid; 0.1 M nitric acid; 1 M sodium hydroxide; 0.5 mM diphenylamine (DPA); 1 mM potassium chloride.
5. Electropolymerization solution: 0.01 M of 3,4-ethylenedioxythiophene (EDOT) and 0.1 M of magnesium perchlorate ($Mg(ClO_4)_2$) dissolved in acetonitrile.
6. Ultrasonic bath.
7. One-compartment electrochemical cell connected to nitrogen line.
8. Reference electrode: silver/silver chloride/potassium chloride (Ag/AgCl/KCl (3 M)).
9. Auxiliary electrode: platinum (Pt) or carbon glass.
10. High-molecular-weight polyvinyl chloride (PVC).
11. Plasticizer (e.g., 2-nitrophenyl octyl ether).
12. Lipophilic salt (e.g., potassium tetrakis(4-chlorophenyl) borate).
13. Freshly distilled tetrahydrofuran (THF).
14. Flat-bottomed glass recipients with ca. 2 cm diameter, puncher with diameter corresponding to that of the electrode body.
15. Philips IS-560 electrode bodies or, alternatively, PVC tubes with diameter ca. 2–3 mm and ca. 10 cm long and Ag/AgCl wire.
16. Potentiostat/galvanostat.

2.4 Potentiometric and Amperometric Measurements

1. Voltmeter with high input impedance (min. 10 MΩ).
2. Galvanostat/potentiostat.
3. Measuring cell.
4. Magnetic stirrer.
5. Volumetric pipette.
6. Reference electrode: Ag/AgCl/KCl (3 M).
7. Auxiliary electrode: platinum or carbon glass.
8. Solutions: 5 mM DPA; 0.1 M KNO_3; 0.1 M HNO_3; 1 M NaOH.
9. Methanol with 10% v/v of acetic acid and water.

3 Methods

3.1 Synthesis of MIP Microparticles

1. Assemble polymerization setup according to Fig. 1 (*see* **Note 1**).
2. Place solution of 1.5 mmol of monomer (MAA) and template in 40 mL of acetonitrile into the reactor. Stir and let equilibrate for at least 15 min.

3. Add 2 mmol of cross-linker and 0.5 mmol of initiator and degas with nitrogen for 15 min.
4. Turn on the condenser water cooler and the water bath pump.
5. Increase water bath temperature to 60 °C and maintain it for the duration of the reaction (10–24 h). Turn off stirring and nitrogen. Reactor side arm for the introduction of reagents should be opened only for adding reagents.
6. After the end of the reaction, turn off heat and let the apparatus cool to room temperature (20 °C). Then, turn off the water cooler and remove polymerization mixture with MIP particles from the reactor.
7. Separate polymer microparticles from the solvent by centrifugation at ca. 8500 × *g*; wash particles with methanol and repeat centrifugation.
8. Assemble Soxhlet setup.
9. Place MIP particles into the extraction thimble and load thimble into the main Soxhlet chamber.
10. Place 60–80 mL (3–4 times the main chamber volume) of methanol/acetic acid mixture in the distillation flask together with some boiling chips.
11. Turn on heater and water for the condenser.
12. Extract template from MIP particles in Soxhlet over a period of ca. 12 h; after that turn off the heater and let the apparatus to cool to room temperature.
13. Remove polymer particles from the thimble and dry them at 50 °C in the laboratory oven until they reach constant weight.
14. Keep MIP particles at room temperature until further use.

3.2 Electrode Preparation with Plasticized PVC Membranes

1. To prepare membrane cocktail mix, first disperse 5 mg of MIP microparticles in 30 mg of plasticizer.
2. Next, add 2.5 mg of lipophilic salt and 15 mg PVC, mix thoroughly, and add ca. 1.5 mL tetrahydrofuran.
3. Homogenize mixture by stirring until microparticles are well dispersed and all other compounds are dissolved.
4. Pour into the flat-bottomed recipient with diameter of ca. 2.5 cm and let solvent evaporate for 24 h. After evaporation, a flexible membrane is formed with a thickness of ca. 0.2 mm.
5. Remove membrane from recipient and cut a disk with a diameter of ca. 7 mm using a punch.
6. Mount cut membrane in the Philips electrode body by fixing and sealing it with silicon O-ring in the screw cap connected to the body, with internal reference electrode built in. Alternatively, cut a disk with diameter corresponding to

the diameter of the PVC tube, glue it to the tube using PVC-cyclohexanone slurry, and let dry. Insert Ag/AgCl wire into the PVC tube.

7. Fill both types of electrodes with enough inner solution, 1 mM KCl, to cover internal reference electrodes.

3.3 Electrode Preparation with Electropolymerized Conducting Polymer Membranes

1. Clean and activate electrode surface prior to the MIP immobilization on the surface. For this, polish electrode surface with 4000-grit abrasive paper followed with a 6 μm diamond polish. Thoroughly wash with ethanol and ultrapure water and dry.
2. For chemical cleaning of film electrodes brush their surface by wetting it, successively, in ethanol, water, sulfuric acid (6 M), and deionized water, and finally dry it under nitrogen flow.
3. For electrochemical activation of film electrodes, assemble the three-electrode electrochemical cell comprised of the reference electrode (Ag/AgCl/KCl (3 M)), auxiliary electrode (Pt or carbon glass), and working electrode (*see* **Note 2** and Fig. 2).
4. Cycle the potential 20 times between 0.8 and −2.2 V in 0.1 M KNO_3 solution, then wash working electrode with water, and dry.
5. Add 10 mg of MIP microparticles to 5 mL of the solution of 0.01 M of EDOT and 0.1 M of $Mg(ClO_4)_2$ in acetonitrile, and disperse by sonication for 10 min. Proceed to the electropolymerization immediately.

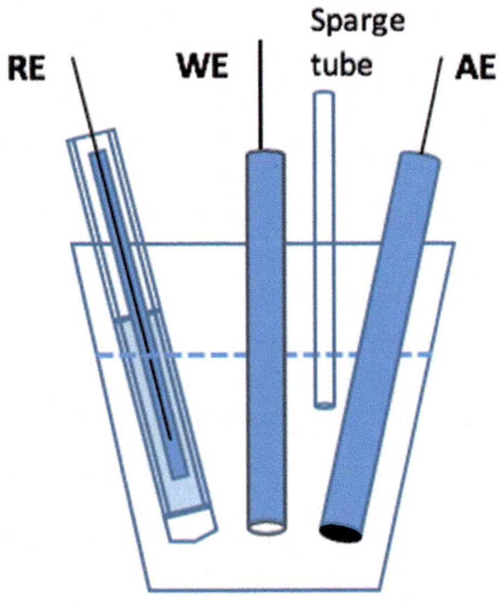

Fig. 2 Schematic of the three-electrode electrochemical cell, *RE* reference electrode, *WE* working electrode, *AE* auxiliary electrode

6. Place electropolymerization solution in the electrochemical cell, turn on nitrogen flow, and sparge solution for 5 min.
7. Perform electropolymerization at room temperature under nitrogen flow (*see* **Note 3**).
   (a) For the preparation of potentiometric sensor, perform electropolymerization at potentiostatic conditions at the potential of 1.2 V until 15 mC charge has been passed. This ensures deposition of the conducting polymer layer containing MIP particles with a thickness of ca. 800 nm.
   (b) For the preparation of amperometric sensor, perform electropolymerization at galvanostatic conditions at the current of 20 μA for 1 min. This ensures deposition of the conducting polymer layer containing MIP particles with thickness of ca. 50 nm.
8. After electropolymerization, rinse electrodes with acetonitrile, dry, and keep them in the air at room temperature until use.

3.4 Calibration Measurements with Potentiometric Sensors with MIP-Modified Membrane

1. Before calibration measurement, condition working sensor in the 0.1 mM solution of analyte (e.g., DFA) overnight and then wash with copious amounts of water until stable potential readings are obtained.
2. For potentiometric measurements, assemble the following galvanic cell comprised of the reference electrode, analyte solution, and working electrode with MIP-modified membrane:

   3 M Ag/AgCl/KCl|analyte solution|sensor.
3. Measure electromotive force (emf) of the cell using high-input-impedance voltmeter.
4. For calibration measurements, place an exact volume of supporting electrolyte, 0.1 M KNO_3, measured using volumetric pipette, in the measuring cell, and turn on stirring.
5. Carry out calibration by adding known amounts of the analyte stock solution, wait for the sensor potential to stabilize (typically ca. 5 min), record it, and repeat the analyte addition and reading process until reaching the limit of linearity.
6. For the application in Subheading 3.6, DPA stock solution with concentration 5 mM was used to prepare calibration solution in the concentration range from 0.1 μM to 0.5 mM.
7. After measurements, wash sensor with copious amounts of water. Store sensors in water between measurements.
8. Plot sensor response in coordinates E (emf of the galvanic cell or sensor potential) versus $log\ a_i$ (logarithm of activity of the measured ion) and calculate parameters of the Nernst equation, i.e., slope of the electrode function, S, and standard potential of the sensor, E^0, using linear regression:

$$E = E^0 + 2.303\frac{RT}{z_i F}\log a_i = E^0 + S\log a_i \qquad (1)$$

where E is the emf of the galvanic cell, E^0 is the sensor standard potential, R is the universal gas constant, T is the absolute temperature (K), F is the Faraday constant, z_i is the charge of the measured ion, a_i is the activity of the measured ion (i), and S is the slope of the electrode function.

The theoretical slope of the electrode function for single charged ion i has a value of 59.18 mV/log a at 25 °C. Activity is related to the concentration through an expression, $a_i = \gamma C_i$, where γ is activity coefficient and C_i is concentration. If ionic strength of measured solutions is kept constant throughout the calibration, i.e., by using supporting electrolyte or ionic strength adjustment buffer, activity coefficients can be considered constant as well and concentration of analyte ion can be used for calculating calibration curve parameters.

3.5 Calibration Measurements with Amperometric Sensors with MIP-Modified Membrane

1. For amperometric measurements, assemble the three-electrode electrochemical cell as described in Subheading 3.3.
2. Prepare analyte calibration solutions as described above in the desired concentration range (e.g., 5–115 μM of DPA).
3. Use HNO_3 solutions with concentration of 0.1 M with pH adjusted to 2 by addition of necessary amount of 1 M NaOH solution as a supporting electrolyte.
4. Turn on nitrogen flow and purge solution for 5 min. Carry out detection of analyte (e.g., DPA) by differential pulse voltammetry (DPV) using the following settings: potential range from 0.6 to 0.85 V; scan rate 0.1 V s^{-1}; standby potential −0.2 V; modulation amplitude 0.025 V; modulation time 0.05 s; and equilibration time 10 s.
5. After calibration, wash electrodes with solution of 1:9 methanol to acetic acid followed by water, and store dry between measurements. Amperometric sensors do not need conditioning prior to measurements.
6. Plot sensor response (area of the recorded current peak) versus analyte concentration and calculate parameters of the calibration curve (offset and slope) using linear regression.

3.6 Determination of DPA in Apple Juice Using Amperometric Sensor with MIP-Modified Membrane (See Note 4)

1. Assemble three-electrode electrochemical cell as described in Subheading 3.3.
2. Filter apple juice, dilute it twofold with 0.2 M HNO_3 solution, and adjust pH to 2 with 1 M NaOH solution.
3. Place diluted apple juice in the cell, turn on nitrogen flow, and purge solution for 5 min.

4. Carry out measurements by differential pulse voltammetry (DPV) with the settings described in Subheading 3.5.
5. Quantify the amount of DPA present in the sample using single addition method. For this, measure sensor response in the diluted apple juice and after adding a small volume of DPA standard solution.
6. Concentration of analyte after addition should be at least two times higher than initial concentration but within linearity range. Calculate the concentration of analyte in the sample using the following formula:

$$C_i = {A_i C_s \frac{V_s}{V}} \Big/ {A_s - A_i \frac{V_i}{V}}$$

where C_i is the unknown analyte concentrations, C_s is the addition (spike) concentration, A_i is the sensor response in the samples before addition, A_s is the sensor response after addition, and V_s/V and V_i/V are the dilution factors, where V_i is the initial sample volume, V_s is the added volume, and V is the final volume (sum of V_i and V_s).

4 Notes

1. The described MIP composition was used for imprinting with DPA [10, 11]; however, it can also be used for other templates. MIP of different compositions can be prepared employing other monomers, cross-linkers, initiators, and solvents. The ratio between solvent and polymer constituents needs to be kept such that the precipitation/polymerization process leads to formation of imprinted polymeric microparticles that can be easily incorporated in the sensor membrane. Spherical MIP particles with diameters ranging from 0.2 to 0.8 μm, depending on the conditions, were obtained. Particle size distribution was homogeneous for each condition. Use of a different radical initiator implies setting a different polymerization temperature, which should be selected according to the initiator half-life (10 h).
2. Other reference electrodes, e.g., a calomel electrode, can be used as well. In this case, adjust potential accordingly, e.g., the difference between the potential of Ag/AgCl/KCl (3 M) and saturated calomel electrode is -0.032 V.
3. Membrane thickness should be different for potentiometric and amperometric sensors due to the difference in their mode of function. In potentiometric sensors, the response-generating process (ion exchange) takes place at the interface between sensing membrane and solution, which makes

membrane thickness relatively irrelevant. However, a thicker membrane is preferable as it ensures long-term stability of the sensor, although it should not be too thick, in order to avoid poor adhesion and peeling. In amperometric sensors, the analyte needs to diffuse to the electrode surface, and MIP serves to pre-concentrate the analyte. Thus, the sensing membrane needs to be thinner than that of the potentiometric membrane to ensure faster diffusion and fast sensor response.

4. Application of potentiometric and amperometric sensors with MIP-modified membranes to the detection of DPA is described as an example. Analytical procedures are generic and can be adapted to other analytes by changing concentration ranges and pH and composition of supporting electrolyte.

Acknowledgments

The authors thank CESAM (UID/AMB/50017—POCI-01-0145-FEDER-007638), FCT/MCTES through national funds (PIDDAC) for financial support, as well as FEDER for providing co-funding within the PT2020 Partnership Agreement and Compete 2020. A. Rudnitskaya wishes to acknowledge postdoctoral fellowship SFRH/BPD/104265/2014 supported by FCT.

References

1. Lokman U, Turner APF (2016) Molecularly-imprinted polymer sensors: realizing their potential. Biosens Bioelectron 76:131–144
2. Cervini P, Gomes Cavalheiro ET (2012) Strategies for preparation of molecularly imprinted polymers modified electrodes and their application in electroanalysis: a review. Anal Lett 45:297–313
3. Delaney TL, Zimin D, Rahm M, Weiss D, Wolfbeis OS, Mirsky VM (2007) Capacitive detection in ultrathin chemosensors prepared by molecularly imprinted grafting photopolymerization. Anal Chem 79:3220–3225
4. Wang T, Shannon C (2011) Electrochemical sensors based on molecularly imprinted polymers grafted onto gold electrodes using click chemistry. Anal Chim Acta 708:37–43
5. Yoshimi Y, Sato K, Ohshima M, Piletska E (2013) Application of the 'gate effect' of a molecularly imprinted polymer grafted on an electrode for the real-time sensing of heparin in blood. Analyst 138:5121–5128
6. Xiao N, Deng J, Cheng J, Ju S, Zhao H, Xie J, Qian D, He J (2016) Carbon paste electrode modified with duplex molecularly imprinted polymer hybrid film for metronidazole detection. Biosens Bioelectron 81:54–60
7. Lia T-J, Chen P-Y, Nien P-C, Lin C-Y, Vittal R, Ling T-R, Ho K-C (2012) Preparation of a novel molecularly imprinted polymer by the sol-gel process for sensing creatinine. Anal Chim Acta 711:83–90
8. Marx S, Zaltsman A, Turyan I, Mandler D (2004) Parathion sensor based on molecularly imprinted sol-gel films. Anal Chem 76:120–126
9. Cennamo N, D'Agostino G, Pesavento M, Zenni L (2014) High selectivity and sensitivity sensor based on MIP and SPR in tapered plastic optical fibers for the detection of l-nicotine. Sensors Actuators B Chem 191:529–536
10. Granado VLV, Rudnitskaya A, Oliveira JABP, Gomes MTSR (2012) Design of molecularly imprinted polymers for diphenylamine sensing. Talanta 94:133–139
11. Granado VLV, Gutierrez-Capitan M, Fernández-Sánchez C, Gomes MT, Rudnitskaya A, Jimenez-Jorquera C (2014) Thin-film electrochemical sensor for diphenylamine detection

using molecularly imprinted polymers. Anal Chim Acta 809:141–147

12. Blanco-Lopez MC, Gutierrez-Fernandez S, Lobo-Castanon MJ, Miranda-Ordieres AJ, Tunon-Blanco P (2004) Electrochemical sensing with electrodes modified with molecularly imprinted polymer films. Anal Bioanal Chem 378:1922–1928
13. Ho K-C, Ye W-E, Tung T-S, Liao J-Y (2005) Amperometric detection of morphine based on poly(3,4-ethylenedioxythiophene) immobilized molecularly imprinted polymer particles prepared by precipitation polymerization. Anal Chim Acta 542:90–96
14. Poma A, Turner APF, Piletsky SA (2010) Advances in the manufacture of MIP nanoparticles. Trends Biotechnol 28:629–637
15. Yoshimatsu K, Reimhult K, Krozer A, Mosbach K, Sode K, Ye L (2007) Uniform molecularly imprinted microspheres and nanoparticles prepared by precipitation polymerization: The control of particle size suitable for different analytical applications. Anal Chim Acta 584:112–121

Chapter 13

Scalable Arrays of Chemical Vapor Sensors Based on DNA-Decorated Graphene

Jinglei Ping and A. T. Charlie Johnson

Abstract

Arrays of DNA-functionalized graphene field-effect transistors (gFETs) hold great promise for high-performance vapor sensing. In this chapter, we describe methods for the scalable production of gFET-based vapor sensors with high sensitivity and efficiency in size, cost, and time. Large-area graphene sheets were prepared via chemical vapor deposition (CVD); a standard photolithographic processing for large-area graphene was used to fabricate gFETs with high mobility and low doping level under ambient conditions. The gFETs were functionalized by single-stranded DNA (ssDNA), which binds to the graphene channels through π–π stacking interaction and provides affinity to a wide range of chemical vapors. The resulting sensing arrays demonstrate detection of target vapor molecules down to parts-per-million concentrations with high selectivity among analytes with high chemical similarity including a series of carboxylic acids and structural isomers of carboxylic acids and pinene.

Key words Vapor sensor, Graphene, Field-effect transistor, DNA

1 Introduction

There is an increasing push to develop sensitive and selective chemical vapor sensors for environmental monitoring, breath analysis and diagnostics, and industrial processing monitoring, among other applications. Carbon-based nanomaterials, including carbon nanotubes and graphene, feature outstandingly high biocompatibility, biostability, and surface-to-volume ratios, ideal for high-sensitivity sensing applications [1]. Implementation of chemical/biosensing applications requires scalable fabrication techniques that are compatible with existing technologies. Graphene, in particular, is readily synthesized over large areas via chemical vapor deposition (CVD) [2], and its planar format enables scalable production of all-electronic chemical sensors [3].

We developed a scalable production of chemical vapor sensors based on DNA-functionalized graphene field-effect transistors (gFETs). Large-area graphene was synthesized via CVD and

Jessica E. Fitzgerald and Hicham Fenniri (eds.), *Biomimetic Sensing: Methods and Protocols*, Methods in Molecular Biology, vol. 2027, https://doi.org/10.1007/978-1-4939-9616-2_13, © Springer Science+Business Media, LLC, part of Springer Nature 2019

transferred onto SiO_2/Si wafers with an electrolytic bubbling method [4]. The gFET arrays were manufactured using photolithography with a sacrificial metal passivation layer deposited on the graphene to avoid potential chemical contamination associated with photolithographic processing. The resulting gFETs were highly reproducible with low doping levels and high mobility (1640 ± 250 cm^2 V^{-1} s^{-1}).

The gFETs were functionalized with single-stranded DNA sequences of various types to detect volatile organic compounds (VOCs), e.g., homologous carboxylic acids and pinene isomers. This functionalization step transforms gFETs into sensors that respond, depending on the sequences of DNA, to the concentration of vapor analytes. The resulting sensors were demonstrated to detect the anaytes at 1 part per million (ppm) with high reproducibility and rapid response (seconds).

2 Materials

2.1 Growth and Transferring of Graphene

1. Copper foil: 0.025 mm (0.001 in.), annealed, uncoated, 99.8% (metal basis).
2. Methane: Ultrahigh purity (4.0 grade, 99.99% CH_4, 5 ppm O_2, 5 ppm H_2O, 20 ppm C_2H_6, <20 ppm other hydrocarbon, 70 ppm N^2).
3. Hydrogen: Ultrahigh purity (5.0 grade, 99.999% H_2, 1 ppm O_2, 2 ppm H_2O, 0.5 ppm tetrahydrocannabinol, <1 ppm CO and CO_2, 5 ppm N^2).
4. Argon: Ultrahigh purity (5.0 grade, 99.999% Ar, 1 ppm O_2, 1 ppm H_2O, 0.5 ppm tetrahydrocannabinol, <1 ppm CO and CO_2, 5 ppm N^2).
5. Nitric acid: 70.0% w/w.
6. Poly(methyl methacrylate) (*see* **Note 1**): 950PMMA A 4.
7. Sodium hydroxide: ≥97.0%, pellets.
8. Polyethylene terephthalate sheet: Adorox, 0.017″ thick.

2.2 Fabrication of gFET Arrays

1. SiO_2/Si substrate: Silicon Valley Microelectronics, 100 mm diameter, p-type boron-doped, <1-0-0>, 1-100 ohm-cm, 525 ± 25 μm, front surface polished, back surface etched, 300 ± 15 nm thermal oxide.
2. Photoresist polymethylglutarimide (PMGI): MicroChem Corp. (PMGI SF).
3. Photoresist S1813: Shipley (Microposit S1813).
4. Developer MF-319: Shipley (Microposit MF-319).
5. Remover 1165: Dow (Microposit Remover 1165).

6. Gold Etchant trifluoroacetic acid (TFA): Transene (Gold Etchant TFA).
7. Acetone: Fisher Scientific (A18P, ≥ 99.5%).
8. 2-Propanol (IPA): Fisher Scientific (A451, ≥ 99.9%).

2.3 Functionalization of gFET Arrays by DNA

1. DNA oligos: Fisher Scientific (Invitrogen Life Sciences, desalted, dry, 5′-end aminated).

2.4 Vapor Sensing Using DNA/gFET Arrays

1. Clean air: Praxair (UN1002), compressed.

3 Methods

3.1 Growth of Graphene via CVD

Monolayer graphene of large area (up to 100 mm × 300 mm) was grown via low-pressure chemical vapor deposition on a copper foil substrate in a 4-in. quartz tube furnace, as shown by the schematics in Fig. 1a.

1. First clean copper chemically by soaking in 1% nitric acid for 30 s, and then thoroughly rinse by deionized (DI) water.
2. Anneal the copper for 60 min at 1020 °C in ultrahigh-purity (UHP) hydrogen (flow rate 80 sccm, pressure 650 mT).
3. Deposit graphene on the copper substrate using UHP methane as the precursor for 20 min (temperature 1020 °C; hydrogen flow rate 80 sccm; methane flow rate 10 sccm; pressure 850 mT).

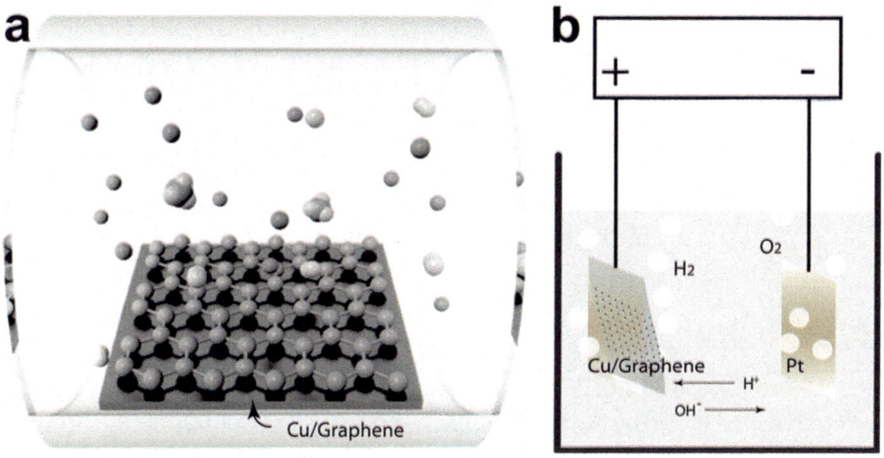

Fig. 1 (**a**) Schematic showing the growth of graphene via chemical vapor deposition in a quartz tube. (**b**) Bubbling transfer method where graphene is mechanically "peeled" from copper substrate by H_2 gas generated in an electrolysis process

4. After deposition, a fast-cooling process is crucial for the formation of high-quality monolayer graphene. To this end, immediately turn off the power supply of the furnace, and then slide the heating region of the furnace along the tube so that the area enclosing the sample is exposed to ambient air. This causes the temperature of the heated area to drop from 1020 °C by 5 °C/s initially, and finally to room temperature within 35 min.
5. Turn off methane flow when the temperature is 650 °C.
6. When the tube reaches room temperature, turn off hydrogen, and follow by flushing with argon for 5 min before the sealing of the quartz tube is removed.

3.2 Electrolytic "Bubble" Transfer of Graphene

Graphene deposited on the copper substrate was transferred onto a silicon wafer with 300 nm dielectric SiO_2 layer using an electrolytic "bubble" method, as shown in Fig. 1b.

1. First, spin a layer of poly(methyl methacrylate) (PMMA), on the graphene/foil (2000 rpm for 1 min).
2. Anneal on hot plate at 150 °C for 10 s.
3. Connect the PMMA/graphene/Cu film to the anode of a DC voltage (20 V).
4. Place a piece of platinum foil in a 100 mM NaOH solution as the cathode, and immerse the PMMA/graphene/Cu film in this solution. In the electrolysis process, hydrogen bubbles are generated between the PMMA/graphene and the copper foil, which mechanically separates the PMMA/graphene layer from the growth substrate. After being displaced from the copper foil, the PMMA/graphene film should float on the surface of the NaOH solution with graphene as the underside.
5. Use a rigid polyethylene terephthalate sheet to lift the floating PMMA/graphene film out of the NaOH solution and then place it into three DI water baths in series to thoroughly wash off ionic residues.
6. After the last water bath, transfer the sample to the target Si/SiO_2 wafer and blow dry gently with compressed nitrogen gas.
7. Gently bake at 100 °C for 2 min to smooth the PMMA/graphene film on the wafer and increase its adhesion to the surface.
8. In order to remove the PMMA layer to leave clean graphene on the substrate, place the PMMA/graphene/substrate in an acetone bath for 30 min, followed by 30 s in spraying acetone and 30 s spraying 2-propanol (IPA).
9. Gently blow dry with nitrogen.

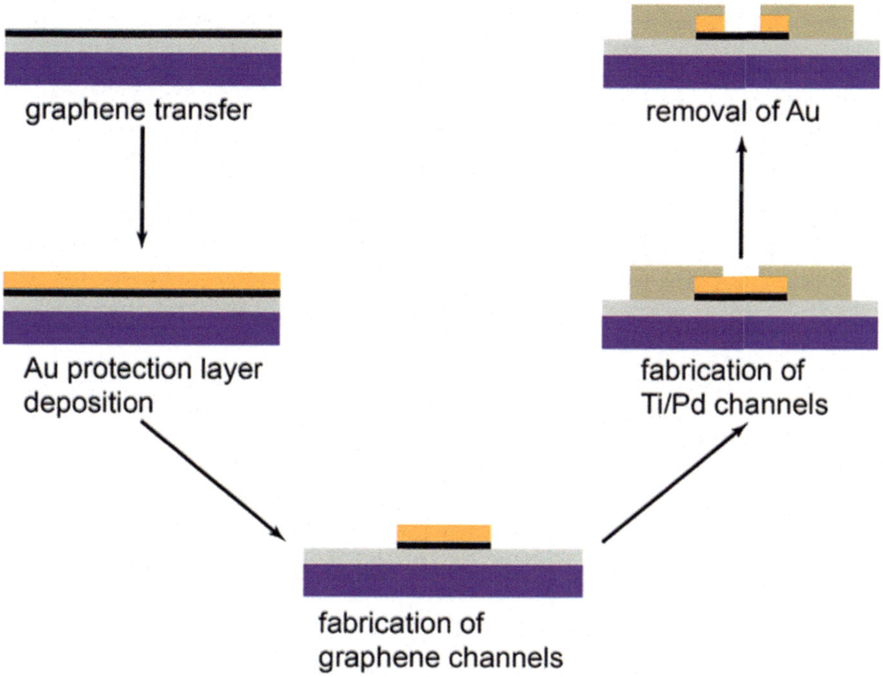

Fig. 2 Scalable micro-manufacturing of field-effect transistors based on pristine graphene using gold protection layer

10. Anneal the resulting graphene/substrate chip in a quartz tube at 200 °C in argon (1000 sccm) and hydrogen (250 sccm) atmosphere for 30 min.

3.3 Scalable Microfabrication of gFET Arrays

The all-photolithographic microfabrication process we developed, shown in Fig. 2, ensures that the gFET arrays and gFET-based vapor sensors are produced in a scalable way.

1. After the cleaning procedure for graphene transferred onto the SiO_2/Si substrate, thermally deposit a 20 nm layer of gold (to protect the graphene from subsequent processing) with a low deposition rate of <0.5 Å/s to ensure best gold coverage quality.

   A bilayer photolithographic process is then used to pattern the graphene channel region.
2. Spin coat a layer of positive photoresist PMGI onto the chip at 4000 rpm for 45 s and bake at 200 °C for 5 min.
3. Next, spin coat a positive photoresist S1813 layer at 5000 rpm and bake at 100 °C for 2 min.
4. Use a mask aligner to expose the chip sample to UV light with wavelength 315 nm and power density 80 mL/cm^2 in soft contact mode (*see* **Note 2**).
5. Place the chip in a developer MF-319 bath and puff gently with a pipette for ca. 30 s until no photoresist can be further

removed. Graphene regions that eventually become the gFET channels remain covered by photoresist.

6. Remove the gold in the exposed regions by wet etching in 1:9 gold etchant TFA:deionized water for 2 h.
7. Dry etch the exposed graphene by oxygen plasma (300 mTorr, 50 W, 30 s).
8. For liftoff of the remaining photoresist, incubate the chip in photoresist remover 1165 for 2 h, followed by thoroughly rinsing with DI water, leaving graphene channels covered by Au layer.

   A second round of photolithography was used to contact the graphene channels with metal electrodes.
9. Spin coat a PMGI/S1813 bilayer onto the substrate using the same procedure described above.
10. Align a photolithographic mask of the electrode pattern to the Au/graphene channels. The contact regions are defined by exposing the device to UV light and developing the photoresist, also using the same recipe as in last step.
11. Then use thermal evaporation to deposit a first adhesion layer of 5 nm Ti with deposition rate <0.8 Å/s.
12. Without breaking the vacuum, deposit a 45 nm Pd layer with deposition rate ca. 2 Å/s.
13. Use 1165 for the liftoff process (**step 8**) to remove the remaining photoresist and metal deposited on top of it, leaving only the contact region (10 × 10 μm channels).
14. Etch the gold protection layer on the graphene channels by using diluted TFA as described above, leaving pristine graphene with Au/Ti/Pd contacts. This method ensured that the graphene channel was free of photoresist residue contamination.
15. Anneal the resulting chip in an open-tube furnace at 200 °C for 1 h in an environment of Ar (1000 sccm) and H_2 (250 sccm).
16. Finally, soak the annealed chip in acetone for 5 h, and thoroughly rinse with IPA for 30 s.

3.4 DNA Functionalization

The functionalization of DNA on the pristine graphene channels transforms the gFETs into vapor sensors (*see* **Note 3**).

1. Dissolve desalted crystals of single-stranded DNA (ssDNA; obtained commercially) in DI water and dilute to a concentration of 100 μM.
2. Pipette droplets of the DNA solution to just-cleaned array of gFETs and incubate in a humid atmosphere for 30 min.
3. Finally, gently blow the droplet off the wafer with pressurized nitrogen.

3.5 Measurement of DNA/gFET Vapor Sensor Response

1. Place the DNA/gFET sensors in a chamber (6 cm × 2.5 cm × 1 cm) and contact the gold electrodes with electrical feedthroughs (*see* **Note 4**).
2. Vapors are delivered into the chamber via a home-built programmable vapor-handling system as follows: At first flow clean air (1–500 sccm) through a bubbler that contains the liquid analyte to create saturated vapor stream of the analyte in clean air; simultaneously, flow a second stream through a bubbler containing water to create a stream of saturated water vapor. The two streams need to be first mixed with each other and then with a stream of clean air with adjustable flow rate to maintain the total flow rate at 1500 sccm with 33% relative humidity.
3. Apply a source-drain voltage of 100 mV to the DNA/gFETs and record the currents (ca. μA source-drain current) and readout of each DNA/gFET, storing to the computer hard drive.

3.6 An Example

As an example, gFETs functionalized by DNA (5′ GAG TCT GTG GAG GAG GTA GTC 3′) were exposed to dilute vapors of propionic, butyric, hexanoic, and octanoic acids, and carboxylic acids, which differ only in the length of the carbon chain. Ten devices were measured simultaneously while pulses of dilute analyte vapors at controlled concentrations were introduced into the chamber. The duration of the pulse was typically 200 s, followed by clean-air flushing for 200 s. The gas flow rate (1500 sccm) and relative humidity (33%) are held constant throughout the measurement. The response curve and the recovery to baseline are reproducible both in time and across devices (Fig. 3a). The inset shows that the

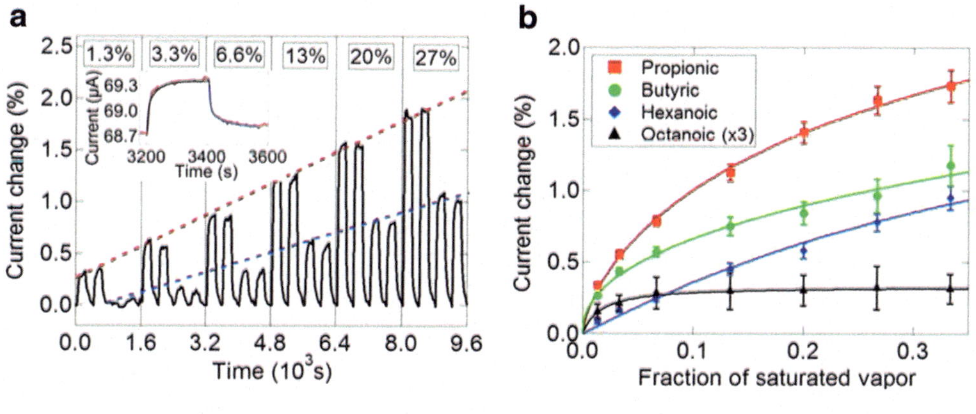

Fig. 3 (**a**) Normalized conductance changes for a DNA/gFET device exposure to propionic acid and hexanoic acid vapor at a variety range of concentrations from 1% to 27% (50–1000 ppm for propionic acid and 3–63 pm for hexanoic acid). At each concentration, two cycles of propionic acid were flushed, followed by two cycles of hexanoic acid. The red (blue) dotted line is used to guide eyes. Inset: fit of the current by double-exponential decays. (**b**) Responses of DNA/gFET as a function of concentration for each of the carboxylic acids tested. Octanoic acid responses have been multiplied by a factor of 3. Error bars represent standard error of the mean of responses based on five devices

response to each pulse can be well fit by double-exponential functions with time constants of 3.5 s (rise) and 4 s (fall). As shown in Fig. 3b, the four carboxylic acids were clearly differentiated at all concentrations tested using the DNA/gFET sensors.

4 Notes

1. All inorganic solutions were prepared in deionized water (DI water).
2. Mask with feature size no less than 5 μm will result in high yield (>90%) of device fabrication based on our recipe.
3. For multiplexed DNA functionalization, special care needs to be taken to avoid cross-contamination of gFETs functionalized with different DNA sequences.
4. Home-built chamber is of size 6 cm × 2.5 cm × 1 cm. Vapors are delivered by using a system of mass flow controller (MKS 1179) with an intermediary controller (MKS 247D). For chemicals of sufficient volume (>10 mL), bubblers with O-ring fittings compatible to ¼″ Teflon tubing are used.

Acknowledgments

This research was supported by the Nano/Bio Interface Center through the National Science Foundation Nanoscale Science and Engineering Center (NSEC) DMR08-32802, and the work involved use of its facilities. Support from Lockheed Martin is also gratefully acknowledged.

References

1. Llobet E (2012) Gas sensors using carbon nanomaterials: a review. Sensors Actuators B Chem 179(31):32–45
2. Li X et al (2009) Large-area synthesis of high-quality and uniform graphene films on copper foils. Science 324(5932):1312–1314
3. Kybert NJ et al (2013) Scalable arrays of chemical vapor sensors based on DNA-decorated graphene. Nano Res 7(1):95–103
4. Gao L et al (2012) Repeated growth and bubbling transfer of graphene with millimetre-size single-crystal grains using platinum. Nat Commun 3:699

Chapter 14

Single-Molecule Mechanochemical Sensing Using DNA Origami Nanostructures

Sagun Jonchhe and Hanbin Mao

Abstract

Single-molecule techniques such as fluorescence-based methods offer superior sensitivity in biosensing. By direct coupling of analyte recognition and signal amplification, a new sensing strategy, single-molecule mechanochemical sensing, has demonstrated high signal-to-noise ratio in the detection of chemicals and biochemicals. However, parallel sensing is limited. In this strategy, DNA origami nanoassemblies with different sensing probes are used as templates for multiplexing tasks. Assisted by the mechanochemical reporting units, these templates are capable of simultaneous detection of biological samples such as platelet-derived growth factor (PDGF) and DNA fragments in microfluidic channels. The origami-based strategy therefore increases multitasking compatibility of the mechanochemical sensing.

Key words Single-molecule technique, Mechanochemical sensing, DNA origami nanostructure, Optical tweezers

1 Introduction

Mechanochemistry is an emerging interdisciplinary field that investigates the coupling between mechanical and chemical processes. Inspired by the principle of mechanochemical coupling, a new sensing mechanism, mechanochemical sensing, has been demonstrated recently [1, 2]. In this sensing strategy, mechanical signal is used to report the binding of a target either at the bulk or the single-molecule level. While single-molecule sensing such as fluorescence-based method has been demonstrated recently with high sensitivity [3], mechanical signatures provide a signal-to-noise level superior to this and other methods. One drawback of the mechanochemical sensing is its relatively low multitasking compability. In a typical optical tweezers or atomic force microscope (AFM)-based setting, only one sensing template can be monitored for the target binding. To increase the throughput of the mechanochemical sensing, either multiple templates or multidimensional templates should be used. In the first approach, magnetic tweezers [4] can be used to tether

Jessica E. Fitzgerald and Hicham Fenniri (eds.), *Biomimetic Sensing: Methods and Protocols*, Methods in Molecular Biology, vol. 2027, https://doi.org/10.1007/978-1-4939-9616-2_14, © Springer Science+Business Media, LLC, part of Springer Nature 2019

multiple sensing templates. This protocol describes the second method to increase the multitasking: using multidimensional sensing templates in which many sensing elements can be incorporated. Herein, we present the protocols to prepare DNA origami nanostructures as multiplexed sensing templates. The DNA origami sensors are first tethered between two optically trapped beads. Binding of a target molecule to the recognition element causes a change in the mechanical signal of the DNA origami template. The sensing experiment is carried out in a microfluidic device to facilitate the administration of different target samples.

2 Materials

2.1 DNA Handle Preparation

1. pET-26b (+) plasmid.
2. Forward primers.
3. Biotin- and digoxigenin-modified reverse PCR primers.
4. SacI enzyme.
5. Spin column.
6. Primers for the biotin-labeled handle:
   (a) Forward primer: 5′-GGC CTC TTC GCT ATT ATG GGA AGG GCG ATC GGT CGG ATT C-X-CGC CGA TCA ACT GGG TGC CAG CGT: $X = O\text{-}(CH_2)_2\text{-}O\text{-}(CH_2)_2\text{-}O\text{-}(CH_2)_2\text{-}O$.
   (b) Reverse biotin-labeled primer: 5′-Biotin-GGG TTC GTG CAC ACA GCC CAG CTT.
7. Primers for the digoxigenin-labeled handle:
   (a) Forward primer: 5′-GGC GCC AGT CAT GGT CAT AGC TGT TAC CGA GCT CGA ATT C -X- CGC CGA TCA ACT GGG TGC CAG CGT: $X = O\text{-}(CH_2)_2\text{-}O\text{-}(CH_2)_2\text{-}O\text{-}(CH_2)_2\text{-}O$.
   (b) Reverse digoxigenin-labeled primer: 5′-Dig-GGG TTC GTG CAC ACA GCC CAG CTT.
8. Bovine serum albumin (BSA).
9. Taq polymerase buffer (10×).
10. Magnesium chloride (25 μM).
11. Deoxyribonucleotide triphosphate (dNTP) mixture (10 mM).
12. Taq polymerase.
13. Agarose.

2.2 DNA Origami

1. M13mp18 DNA.
2. DNA staple sequences [2].

2.3 Analyte

1. Platelet-derived growth factor (PDGF) (R&D Systems, Inc., Minneapolis, MN, USA) reconstituted to make stock solution of 8 μM in 4 mM HCl. Dilute the sample further to desired concentration using origami buffer.
2. Complementary DNA:
   (a) 5′-CCA CAT TGC T TTT TTC AGG TTT CTC TTC GGA ATC ACT TCC AAA.
   (b) The complementary DNA was reconstituted with 1 mM Tris buffer to make 1 mM of a stock solution. The stock was further diluted to desired concentration using origami buffer.

2.4 Microfluidic Chamber

1. 24 × 60 × 0.16 mm (w × l × d) cover glass (VWR).
2. Parafilm (Bemis Co., Neenah, WI).
3. Micropipettes (Garner Glass Co., i.d. 25 μm).
4. Versa laser cutter.
5. Hot plate.

2.5 Origami Buffer for Mechanical Unfolding

1. 10 mM Tris–HCl.
2. 1 mM EDTA.
3. 10 mM $MgCl_2$.
4. 100 mM NaCl with pH adjusted to 7.4.

2.6 Targets in Origami Buffer Preparation

1. Dilute 6.25 μL of PDGF stock solution in quantity-sufficient 1000 μL origami buffer to make 50 nM PDGF target solution.
2. Dilute 5 μL of complementary DNA stock solution in quantity-sufficient 1000 μL origami buffer to make 5 μM target DNA solution.
3. Dilute 3.125 μL and 5 μL of PDGF and complementary DNA stock solutions, respectively, in quantity-sufficient 1000 μL origami buffer to make 25 nM PDGF and 5 μM complementary DNA target solution.

2.7 Beads for Mechanical Unfolding Experiment

1. Anti-digoxigenin-coated polystyrene bead.
2. Streptavidin-coated polystyrene bead (Spherotech, Lake Forest, Illinois, USA).

2.8 Instruments

1. Optical tweezers.
2. Thermocycler.

3 Methods

3.1 Preparation of Digoxigenin/Biotin Handles

1. Take a 1.5 mL Eppendorf tube and add 2.5 μL of BSA (×100).
2. Add 151 μL of filtered water, 25 μL of Taq buffer (×10), 10 μL of magnesium chloride (25 μM), and 10 μL of dNTP mixture (10 mM).
3. Add 10 μL of PET-26b (+) (1 ng/μL).
4. Add 125 picomoles of forward primer and 125 picomoles of reverse primer. (Notice that forward and reverse primers are different for digoxigenin and biotin handles.)
5. Finally add 3 μL of Taq polymerase and mix well. The final volume is 250 μL.
6. Split the sample into 50 μL in each small polymerase chain reaction (PCR) tube.
7. Run the PCR in the thermocycler.
8. Mix all the contents in each PCR tube and run the 1% (w/v) agarose gel. The gel should show a band at 2520 base pair (bp).
9. Use the PCR purification kit to purify the PCR product and run the 1% (w/v) agarose gel again to confirm the purity.

3.2 Preparation of 7-Tiled DNA Origami Construct

1. Mix and anneal 0.01 μM of linear M13mp18 DNA with 0.04 μM of each DNA staple strands from 85 °C to 15 °C at a rate of −1.0 °C per minute in a buffer containing 20 mM Tris–HCl (pH 7.4), 1 mM EDTA, and 10 mM $MgCl_2$.
2. Add and mix 0.04 μM of each of the digoxigenin and streptavidin handles to the above reaction.
3. Run the annealing from 50 °C to 15 °C at a rate of −1.0 °C per minute. Refer to Fig. 1 (*see* **Note 1** for the storage condition).

3.3 4-Channel Microfluidic Chamber Setup

1. Drill four holes at each edge of a cover glass by the Versa laser cutter (*see* **Note 2**).
2. Cut the parafilm by the Versa laser cutter according to a four-channel pattern. Remove the cut portion with the help of tweezers (pattern shown in Fig. 2, white area shows the removed parafilm portion).
3. Arrange the cut parafilm on the top of the cover glass drilled with holes.
4. Place micropipettes to connect the top and bottom channels to the middle-top and middle-bottom channels, respectively.
5. Place a clean, intact cover glass on top of the setup.
6. Sandwich the entire setup between two glass slides, put on the hot plate set at 90 °C, and apply 5 lbs. weight for 20 s.

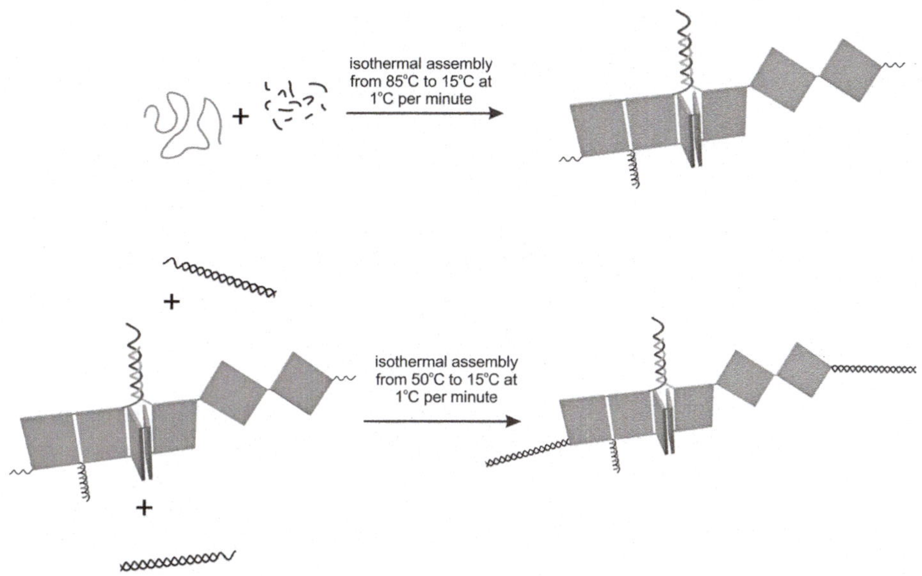

Fig. 1 Synthesis of DNA origami sensing template with two double-stranded DNA handles (*see* Fig. 3 for the attachment of origami to optically trapped beads)

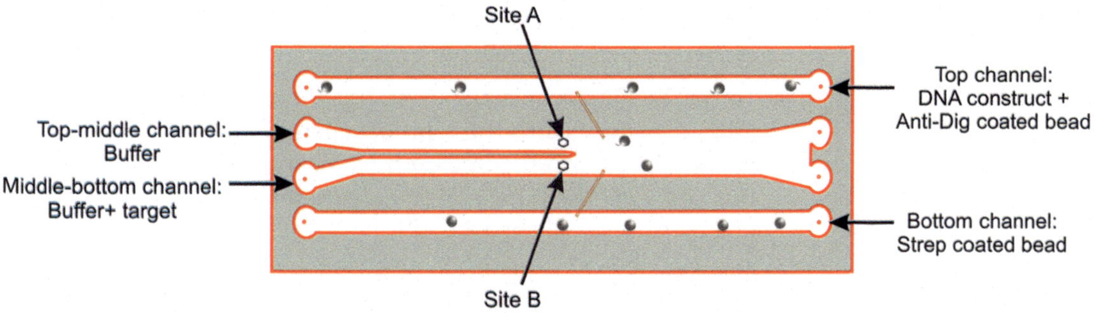

Fig. 2 Flow patterns in the microfluidic setup. "Dig" stands for digoxigenin and "Strep" stands for streptavidin

7. Remove the weight and press the slide with tweezers until the parafilm membrane is transparent, which indicates effective bonding of the parafilm membrane between the two coverslips (*see* **Note 3**).
8. Align the microfluidics chamber in the bracket so that the holes in the chamber match with those of the inlets and outlets of the channels in the bracket (*see* **Note 4**).
9. Flow 70% ethanol into the topmost and bottommost channels. Monitor for any leakage in the middle channels (*see* **Note 5**).
10. Flow the buffer solutions into the microfluidic channels. Monitor for any leakage.

3.4 Setting the Bracket with the Microfluidic Chamber in the Optical Tweezers

1. Adjust the handle of the bracket on a sample stage in the optical tweezers setup so that the microfluidic chamber is placed between two objective lenses of the optical tweezers setup. Add meniscus water between the two chamber surfaces and the two objectives.
2. Adjust the focus of the laser to obtain the diffraction ring patterns monitored by a camera. Adjust the angle of the bracket to obtain symmetric ring patterns on the monitor (*see* **Note 6**).
3. Turn off the laser shutter and locate the positions of the micropipettes connecting the microfluidic channels and the interfacial membrane (*see* **Note 7**).
4. Turn on the laser shutter, focus the trapping laser, and filter out the laser light with a filter wheel so that the laser pattern is no longer shown on the monitor (*see* **Note 8**).

3.5 Mechanical Force Ramping Experiment

1. Incubate 4 μL of the DNA construct with 2 μL of 0.25% (w/v) solution of the anti-digoxigenin (dig)-antibody-coated polystyrene beads in a tapping machine (one tap per second) for 15 min in an Eppendorf tube (*see* **Note 9**).
2. Dilute and disperse the sample in 1000 μL origami buffer.
3. Disperse streptavidin-coated beads in another 1000 μL origami buffer.
4. Inject the DNA construct immobilized on the anti-digoxigenin (anti-dig) beads into the topmost channel and the streptavidin-coated beads into the bottommost channel. Flow buffers without and with the target on the top-middle and bottom-middle channels, respectively (Fig. 2).
5. Capture DNA-containing anti-dig-coated and streptavidin-coated beads in the top-middle and bottom-middle channels, respectively. Move two trapped beads to the place where no target is present (site A in Fig. 2) (*see* **Note 15**).
6. Using a steerable mirror, bring the two trapped beads closer to each other to form the DNA tether in between. The formation of the tether is confirmed by the increase in the force upon increasing the extension between the two beads. The scheme of the DNA tethered between the two optically trapped beads is shown in Fig. 3 (*see* **Note 10**).
7. Start collecting the F-X traces at the loading rate of 5.5 ± 0.5 pN/s using the Nano-MTA controlled steerable mirror which guides one of the optically trapped beads. Note the unlocking (or unfolding) features in the top-middle channel (buffer channel). Two unlocking features are observed with ca. 15 nm and ca. 40 nm change in extension (Δx), which correspond to the unfolding of the origami tiles bound with the PDGF and complementary DNA, respectively.

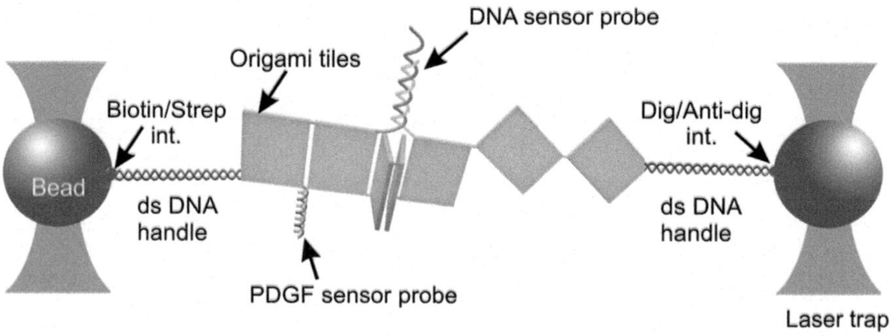

Fig. 3 Schematic setup for the DNA origami-based mechanochemical sensing template

8. Take the tether to the target channel (site B in Fig. 2) and start the data collection using the procedure described in **step 7**. The features disappear once the construct is taken to the buffer channel, which indicates the binding of the targets to the templates. *See* Fig. 4a, c, e for the *F*-*X* curves (*see* **Note 11**).

3.6 Constant Force Experiment

1. Obtain the single-molecule tether, hold the force at 8 pN, and collect the extension versus time plot in the buffer channel (site A of Fig. 2) (*see* **Notes 12** and **13**).
2. During data collection, move the DNA construct to the middle-bottom channel (site B in Fig. 2). The binding of the target is indicated by a sharp change in the extension. *See* Fig. 4g, h for the extension versus time plots (*see* **Note 14**).
3. Once the extension is increased (unlocking of the connection in the origami structure), relax the tether to 0 pN and repeat **steps 1** and **2** for the more data collections.

3.7 Data Analysis

1. Retrieve the data containing *F*-*X* curves using Matlab programs and measure the unlocking force and change in extension (Δx) using Igor programs. *See* Fig. 4a, c, e for the *F*-*X* curves.
2. Plot the Δx histograms for the buffer and target channels. Since unlocking features should not form once the target binds to the probe in the target channel, the counts of the peaks in the Δx histograms decrease in the target channel. *See* Fig. 4b, d, f for the Δx histograms.
3. Plot extension versus time curves using Matlab programs and measure the change in extension (Δx) by Igor programs. *See* Fig. 4g, h for the extension versus time curves.

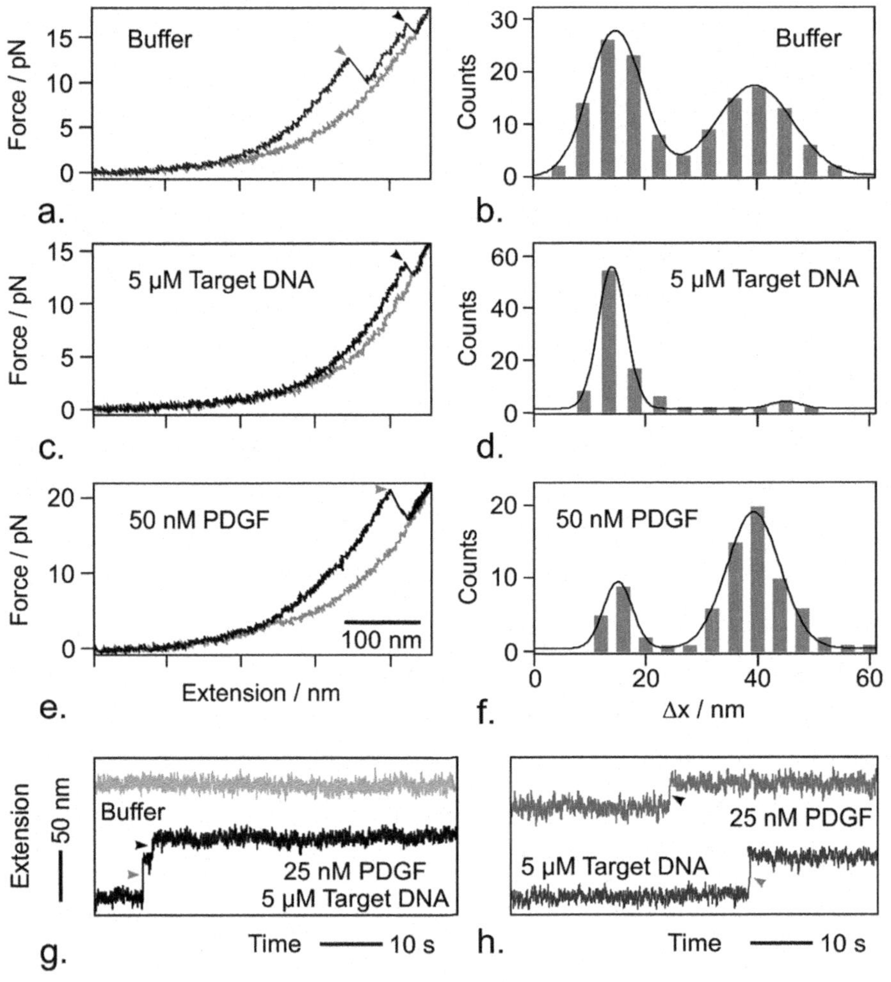

Fig. 4 Data for the mechanochemical origami sensing. (**a**, **c**, **e**) Typical force-extension curves. (**b**, **d**, **f**) The corresponding change-in-extension (Δx) histograms. (**a**, **b**) depict target-free solution; (**c**, **d**) depict 5 μM target DNA; and (**e**, **f**) represent 50 nM PDGF. In the target-free solution, two force-induced unlocking events were observed. In the presence of one target, one unlocking event was observed. (**g**, **h**) Real-time detection of multiple targets in the constant-force mode (8 pN) in a buffer that contains 25 nM PDGF, or 5 μM target DNA, or both. In the target-free solution, the sensor showed no extension jumps. (**g**) When both targets were present, two extension jumps corresponding to the breaking of the two locks were observed. (**h**) In the solution that contains only one target, a corresponding extension jump event was observed. The figure has been rearranged from the Fig. 4 of reference [2].

4 Notes

1. Store the DNA construct at 4 °C.
2. Make sure that the holes in the glass slide are drilled smoothly during the preparation of a microfluidic chamber by the help of a small needle.

3. During the preparation of microfluidic chambers, application of too much pressure may block the micropipette openings due to the melting of the membrane close to the micropipette openings.
4. Make sure that the microfluidic chamber is aligned well with the channel holes in the bracket.
5. While flowing the alcohol in the top- and bottommost channels, overflow of the alcohol into the middle channel through the micropipette is normal, but overflow to other places indicates leaking in the parafilm membrane between different channels.
6. The symmetric diffraction ring pattern indicates that the chamber is symmetrically aligned with the objective lens in the laser tweezers.
7. During the location of the micropipette and the interfacial membrane, laser may cause the formation of air bubbles when the laser is taken close to the micropipette tip or the interfacial membrane.
8. Be sure not to touch the surface of the filter lens.
9. Tapping for longer time can weaken the DNA tether.
10. To confirm the single-molecule tether between the two optically trapped beads: (1) the tether should have an extension of ca. 1700 nm (which is around the contour length of the entire DNA construct) at 31 pN from the point where the two beads touch each other, (2) the tether should have the 65 pN plateau in the *F-X* curves, and (3) when the tether breaks, it should show a single breakage event.
11. Mechanical force ramping experiment is used to find the number of unlocking features.
12. Constant force experiment is used for real-time detection of the target.
13. In the extension versus time curve in the buffer channel, the jumps in the extension are not observed.
14. PDGF in the target channel gives a single-jump event with a change in extension of ca. 15 nm, while the complementary DNA in the target channel yields a single-jump event with a change in extension of ca. 40 nm.
15. During the experiement, watch for the trapping of two beads or any dirt particle in a single laser focus during the experiment. If observed, discard these trapped objects and proceed with a new trapped bead.

Acknowledgments

We thank the NSF (CHE-1609514 and CHE-1415883) for financial support. DNA origami was prepared by Masayuki Endo and Hiroshi Sugiyama at Kyoto University.

References

1. Koirala D, Yu Z, Dhakal S, Mao H (2011) Detection of single nucleotide polymorphism using tension-dependent stochastic behavior of a single-molecule template. J Am Chem Soc 133 (26):9988–9991
2. Koirala D et al (2014) Single-molecule mechanochemical sensing using DNA origami nanostructures. Angew Chem Int Ed Eng 53:8137–8141
3. Tyagi S, Kramer FR (1996) Molecular beacons: probes that fluoresce upon hybridization. Nat Biotechnol 14(3):303–308
4. Assi F, Jenks R, Yang J, Love C, Prentiss M (2002) Massively parallel adhesion and reactivity measurements using simple and inexpensive magnetic tweezers. J Appl Phys 92 (9):5584–5586

Chapter 15

Response Standardization for Drift Correction and Multivariate Calibration Transfer in "Electronic Tongue" Studies

Vitaly Panchuk, Valentin Semenov, Larisa Lvova, Andrey Legin, and Dmitry Kirsanov

Abstract

The procedures for response standardization in "electronic tongue" (ET) studies are described. The construction of reliable multivariate calibration for "electronic tongue" requires the analysis of a large number of representative samples both with ET and reference techniques. This is a laborious and expensive process. Long-term sensor array operation leads to the changes in sensor response characteristics and thus invalidates the multivariate predictive models. Moreover, due to the individual parameters of each sensor in different sensor arrays, it is not possible to use the calibration model for one system together with the data acquired by another system, even if they have the same sensors. Both of these issues lead to the necessity of frequent sensor array calibration which would be ideal to avoid. Instead of recalibration, these two problems can be handled using mathematical methods intended for sensor response standardization. This chapter describes two popular methods of standardization which can be used for both drift correction and calibration transfer. Thus, significant efforts on measuring representative sample sets for sensor array recalibration can be avoided.

Key words Electronic tongue, Multivariate calibration, Response standardization, Calibration transfer, Drift correction

1 Introduction

One of the most widely employed types of sensors in "electronic tongue" (ET) multisensor systems are potentiometric sensors [1, 2]. Due to the physics and chemistry behind the interaction of liquid sample components and sensor membranes, the substantial drifts of potentiometric sensor responses can be observed over a long period of time. This means that response of the same sensor in the same sample will not be identical if measured in different time periods. The responses of the same types of sensors from different production batches can also vary significantly.

Jessica E. Fitzgerald and Hicham Fenniri (eds.), *Biomimetic Sensing: Methods and Protocols*, Methods in Molecular Biology, vol. 2027, https://doi.org/10.1007/978-1-4939-9616-2_15,

Due to the multivariate nature of ET response, the results of measurements with sensor arrays are usually processed with multivariate techniques (e.g., chemometric tools). These techniques take into account the whole response of the multisensor system (the readings from all the sensors) instead of considering each particular sensor separately. The benefits of multivariate data treatment are well known in chemometrics, and application of these tools can often significantly improve the analytical performance of various instruments by enhancing selectivity, improving detection limits, and providing tolerance against noisy measurements, among others [3]. At the same time, appropriate multivariate calibration typically requires many more samples than, for example, ordinary least squares regression with a single response variable. In the latter case, International Union of Pure and Applied Chemistry (IUPAC) recommends using at least six samples with different concentrations [4]. Indeed, the number of calibration samples for a robust multivariate model typically starts from several dozens and, for large-scale industrial applications, can reach many thousands of samples [5].

As the analytical application of ETs is based on multivariate calibrations operating with raw sensor response [6], the sensor drifts will inevitably invalidate the multivariate models. Due to the mentioned inter-batch differences in sensor response, the same calibration model typically cannot be directly applied to several sensor arrays of the same composition (the problem of calibration transfer). The recalibration of a multisensor system seems to be an obvious way to circumvent these issues; however, it can be very cumbersome as it requires a significant number of calibration samples with known characteristics and acquiring and remeasuring them can be quite expensive and tedious. Obviously, it is not practical to perform new calibration measurements for the whole set of samples for each new multisensor system or each time the sensor responses have changed due to the drifts. Another way to handle sensor drifts and calibration transfer between different sensor arrays is to correct sensor responses mathematically. There are various methods for response standardization suggested in literature, mainly for spectroscopic applications [7, 8]. Their application affords the use of multivariate regression or classification model obtained for a particular sensor array together with the data acquired from another sensor array (or the same one with the shifted response) in measuring similar samples. The standardization procedure requires the measurements in a comparatively small standardization set. In spite of the fact that the response standardization is the same from the mathematical point of view for both the drift correction and the calibration transfer, there are certain methodological differences from the analytical chemistry point of view. In case of calibration transfer, these standardization samples (transfer set) are a representative set of calibration samples which uniformly covers the variability in the calibration data. Typically, the transfer set is a small part of the calibration set that has to be measured on the second instrument that is going to be used with the calibration model established earlier for another sensor array [7, 8]. In case of drift correction, the standardization set is much easier—it is just a set of relatively simple standard solutions, which can help to account for sensor drifts.

The standardization methods can be multivariate in nature, i.e., addressing the whole response vector of the instrument with multiple variables simultaneously (the whole spectrum in case of spectroscopic data), or they can be univariate and deal with each particular variable (spectral wavelength in spectroscopy) separately [8]. The choice of particular method for multisensor system depends mainly on sensor type, array configuration, and analytical task. Multivariate standardization, when considering the whole response vector of the system, can be an option when there are many correlated variables in the data, while single sensor standardization is advised for the cases when the neighboring response variables are not well correlated, which is exactly the case for most of the potentiometric multisensor systems [9].

Below, we describe two protocols for response standardization. The first is intended for calibration transfer between ET sensor arrays based on a sensor array response standardization procedure which was initially suggested for multivariate model transfer in spectroscopy as direct standardization [7]. It must be noted that several other methods of transfer are described in literature for multisensor arrays; however, the method detailed below has a benefit of being applicable for transfer between arrays of different sensors; for example calibration derived for potentiometric sensors can be employed to interpret the response of a voltammetric sensor array [10]. Obviously, several important requirements have to be met in order to make this possible, and these are detailed in further sections. The second protocol is intended for drift correction in potentiometric multisensor systems based on standardization of a single sensor response, which is analogous to the single-wavelength standardization in spectroscopy.

2 Calibration Transfer

2.1 Materials

1. Two correctly functioning multisensor systems: They will be further denoted as **1** and **2**, where we assume that system **1** was employed to establish a multivariate calibration model with sufficient number of calibration samples. The data from system **2** are supposed to be employed together with the calibration model derived for the system **1**. As such, system **2** has no separate calibration model.

2. A multivariate calibration model constructed for system **1** with sufficient number of calibration samples that affords desirable precision in prediction: This model can be either regression (for quantification of particular quality parameters or analytes in samples) or classification (for determining sample attribution to the particular sample class).
3. A transfer set of samples—comparatively small (typically less than 10) set of calibration samples employed for the model construction with system **1**, which can be measured also with system **2**: Particular requirements to this set will be detailed below.
4. Software package that is capable of performing linear algebra matrix operations including matrix pseudoinversion.

2.2 Methods

1. Select the transfer set of samples to be measured with multisensor system **2**. The following considerations should be considered:
   (a) The number of samples in the transfer set is typically not fewer than five; the more samples that are employed for transfer, the better the precision in further prediction with the model that can be achieved. However, the price to pay is additional efforts involved with the measurements of these samples with system **2**. Generally, the required number of transfer samples is very task sensitive and may vary depending on the particular nature of the data.
   (b) Transfer samples are selected from the whole calibration set employed for model construction using the data obtained from system **1**. The general requirement for the transfer set is that it has to sufficiently represent variance in the calibration set.
      (i) In the case of regression modeling, this representative coverage may assume the uniform distribution of transfer samples over the whole concentration range.
      (ii) In case of classification models, uniform coverage of all the studied classes in the transfer set has to be met.
      (iii) For multisensor system data, the choice of representative transfer set is not always obvious. There are various ways for subset selection [11]; the most popular one is probably Kennard-Stone algorithm [12]. On the other hand, the use of principal component analysis score plots [13] allows effective visualization of multivariate data structure and these plots can be employed for simple and intuitive transfer set choice.
2. Calculate the transfer coefficients.
   (a) Calculate the transfer coefficients according to the direct standardization procedure [7].

(b) Consider two multisensor systems:

(i) The developed calibration model based on representative set of samples from which we will need to know the system response in the transfer set (X_{1tr}) (*see* **Notes 1** and **2** for transfer set selection considerations).

(ii) The one for which we need to adjust its response in order to use its data together with calibration model established for the first system.

(c) There are two data sets associated with the second system:

(i) Its response in the transfer set measured also with the first system (X_{2tr}).

(ii) Its response in the new samples (X_2).

(d) The calculation of transformation matrix F is done according to

$$F = X_{2tr}^{+} X_{1tr} \quad (1)$$

where + denotes matrix pseudoinverse.

3. Calculate the corrected data for the second multisensor system, X_{2cor} according to

$$X_{2cor} = X_2 F \quad (2)$$

4. Now the calibration model established for the first multisensor system can be employed together with the data derived from the second multisensor system (X_{2cor}).

2.3 An Example for Calibration Transfer

1. Let us consider an example of calibration transfer between two multisensor systems, each having seven potentiometric sensors (*see* **Note 3**), using direct standardization protocol. The response of two multisensor systems comprising the same sensors will typically be different in absolute response values (Fig. 1). The response value of the sensor **S1** is smaller for system **2**, while the response value for **S6** is bigger. In case of potentiometric sensors, this will be due to the minor differences in sensor membrane composition, quality of reagents, and evolution of the sensors upon contact with various media. This difference in response values will lead to the fact that the direct interpretation of the response of the system **2** using the multivariate regression model derived for system **1** will not be possible.

2. System **1** has a calibration model calculated with a sufficient number of calibration samples (Fig. 2, left). Direct application of this model to the data obtained with the second multisensor system will yield inacceptable precision in prediction of the properties of new samples (Fig. 2, right), because the values

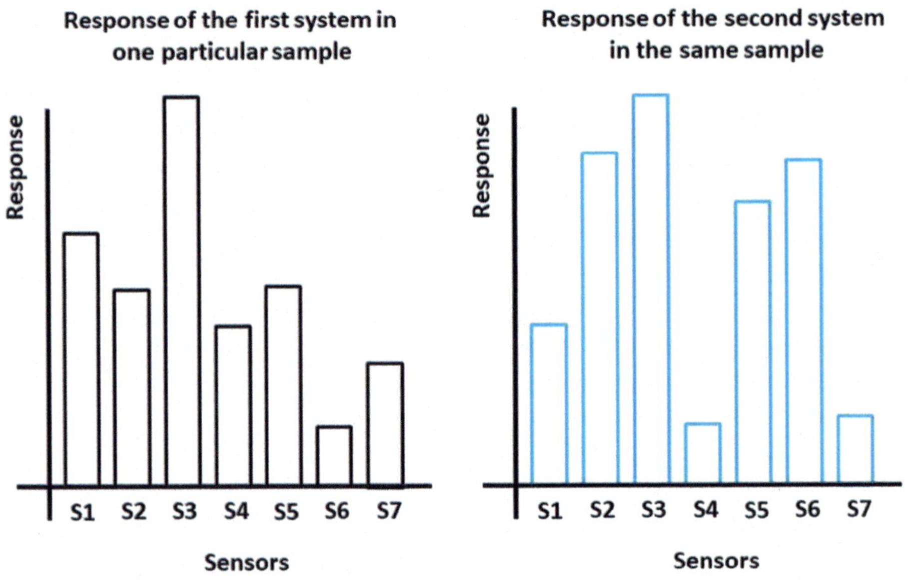

Fig. 1 Responses of two multisensor systems in the same sample

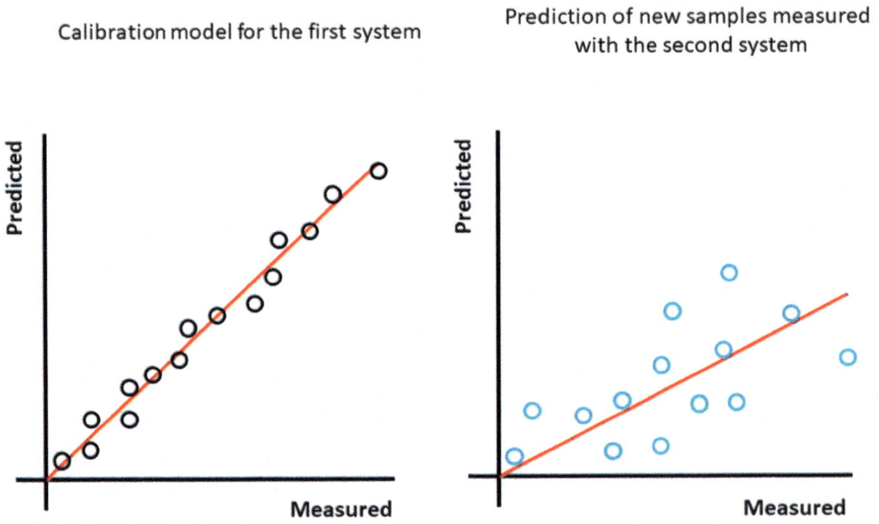

Fig. 2 The use of calibration model with the data from the first and the second multisensor systems

of regression coefficients are associated with particular response values from particular multisensor array.

3. In order to use the first calibration model with the data from the system **2**, direct standardization can be applied. During this procedure, corresponding transformation coefficients will be calculated, which will transform the response of system **2** into the values as if they were measured with system **1**.

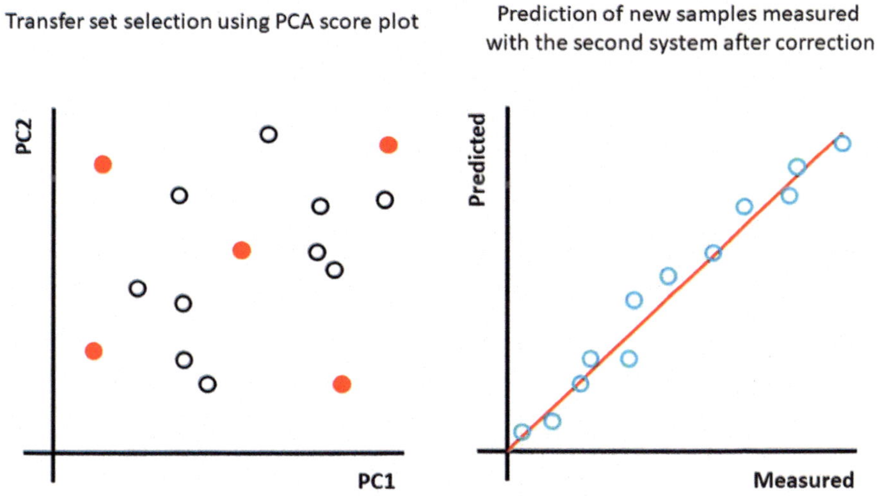

Fig. 3 Transfer set selection and the use of calibration model with the corrected data from the system **2**

4. Calculation of the transformation coefficients requires a proper choice of the transfer set, which is the set of samples measured on both systems. A possible way to choose the transfer set is the use of principal component analysis score plot. Figure 3 (left) shows PCA score plot for calibration samples measured with the first system. In this case, the transfer set samples have to cover the score space in a uniform way in order to account for all variability in the data. The five samples marked with red color in Fig. 3 (left) seem to be an appropriate choice.
5. Using the results of measurements in transfer set with both multisensor systems, the transformation matrix is calculated according to Eq. 1 in Subheading 2.2. This matrix is employed to correct the response of the system **2** and thus afford use of the data from system **2** together with model based on the data from the system **1** (Fig. 3, right).

2.4 Calibration Transfer Notes

1. When choosing the samples for the transfer set, one should also consider their easy accessibility and long-term storage, thus to ensure that the transfer can be performed several times to different multisensor systems.
2. One must also consider the variability of sample matrix when choosing transfer set samples, especially in complex multicomponent media. Thus, the simple uniform distribution of samples over analyte concentration range may sometimes not be sufficient. Generally, the choice of transfer set must obey the same common considerations as the choice of the calibration sample set.
3. The direct standardization procedure can also be employed to transfer calibration models between multisensor systems

employing different types of analytical signal, e.g., between potentiometric and voltammetric [10]. In order to make this unusual procedure possible, both systems should have the response toward target quality employed in calibration. Moreover, the matrix influence on analytical signal for both systems should be comparable.

3 Drift Correction

3.1 Materials

1. Potentiometric multisensor system: correctly functioning multisensor array with any number of sensors. The system should be connected to computer in order to acquire the measurement data in digital format for further processing.
2. Standard solutions: At least three different standard solutions should be employed with the following conditions (*see* **Note 1**):
   (a) The composition of these solutions should cover (if possible) the whole spectra of sensitivity of the sensors in the array. This means that each sensor of the array should have pronounced sensitivity toward at least one of the components of these solutions. If cross-sensitivity of the sensors leads to the response of several sensors toward one component, then there is no need to add extra components to the standard solution.
   (b) The standard solutions should differ either in concentration of the components or in their composition in order to provide for reasonable difference in each sensor reading between these solutions (*see* **Note 2**).
   (c) The concentrations of the components in the standard solutions should not be near the lower or upper detection limit of the sensors but should be in a linear response range of the sensors (typically, this means that components in 10^{-4}–10^{-3} M concentration are added).
   (d) The composition of the standard solutions may differ substantially from that of the target samples (e.g., one can standardize sensor response, for instance, in milk samples using aqueous solutions of several simple inorganic salts).
3. A software package that is capable of performing linear algebra matrix operations or calculating linear regression coefficients.

3.2 Methods

1. Prepare standard solutions in the amount sufficient for the duration of the planned experiment (*see* **Note 3**).
2. Perform the potentiometric measurements with the sensor array in all standard solutions on the first day of the experiment

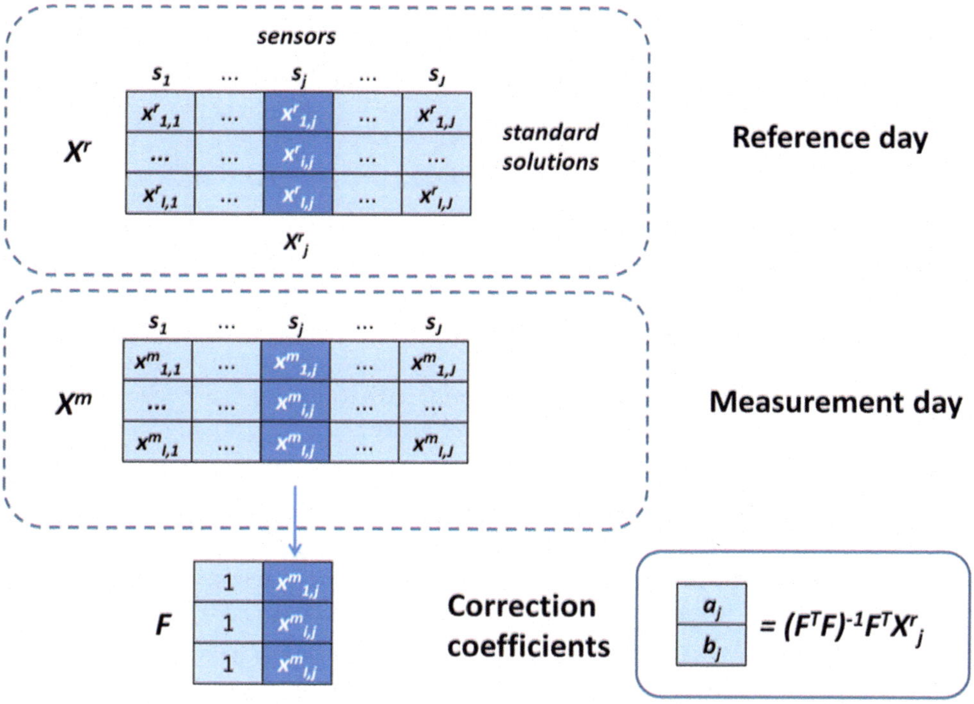

Fig. 4 Data organization for calculation of correction coefficients

and make a record with corresponding results in a file (*see* **Notes 4** and **5**). Later, these measurements will be called "reference day"; they will be designated as matrix X^r, having dimensions $I \times J$, where I is the number of standard solutions, and J is the number of sensors. Each element of this matrix $x^r_{i,j}$ will contain the response of the particular sensor j in a particular standard solution i (*see* Fig. 4 for more details).

3. Perform the measurements in all standard solutions on the second day of the experimental session and make corresponding records with sensor responses in electronic data file (*see* **Note 6**). These measurements will be designated as X^m (Fig. 4), having the same dimensions and structure as X^r.
4. Calculate the correction coefficients a_j and b_j for each particular sensor j on the second experimental day according to the matrix equation in Fig. 4. These coefficients will be employed to adjust the sensor readings as if they were measured on the reference day and thus will eliminate to a certain extent possible sensor drifts occurring between reference and measurement days. This procedure should be repeated for each sensor of the array.
5. The drift in the data acquired from samples of interest measured on the second day now can be corrected using calculated correction coefficients (*see* **Note 7**). The potentiometric

data from samples (matrix D) should be stored in the same format as the data acquired in the standardization solutions. Correct the data D according to the following equation (*see* **Note 8**):

$$D^{cor}{}_{i,j}=a_j+b_jD_{i,j} \tag{3}$$

Here $D^{cor}{}_{i,j}$ is the corrected response of the sensor j in the sample i, $D_{i,j}$ is the raw response of j in the sample i, and a_j and b_j are correction coefficients for the sensor j. This procedure should be repeated for each sensor of the array. D^{cor} constructed from individual $D_{i,j}$ will contain the sensor responses with eliminated drift as if they were measured on the reference day (*see* **Note 9**).

6. Repeat **steps 3–5** for each new day of measurements with the samples of interest: acquire the data in standard solution to receive X^m, perform the measurements in samples to receive D, calculate a and b coefficients for each sensor, and correct the sensor response values in D (*see* **Note 10**).
7. D^{corr} can now be used together with previously obtained multivariate models.

3.3 Drift Correction Example

As an illustrative example below, we are considering the case of multisensor system drift correction in toxicity measurements [14]. In this study sensor array comprised of seven potentiometric sensors was applied for assessment of toxicity induced by microcystin in aqueous samples. The whole experiment was split into several measuring sessions and took over 3 months. The reference data on the toxicity were obtained from microcystin concentration measurements performed with the standard chromatographic technique, ultrahigh-pressure liquid chromatography with diode array detector (UHPLC-DAD). Microcystins were released during the growth of two different strains of *Microcystis aeruginosa* (*M. aeruginosa*) cyanobacteria and the purpose of the study was to distinguish toxin-producing and nontoxic cyanobacteria strains present in potable water.

1. During each measurement session along with potable water samples contaminated with cyanobacteria (toxic and nontoxic), three standard solutions were measured, two of which were mixtures of inorganic salts in the buffer media:
   (a) Solution #1—$CaCl_2$ 0.01 mol/L; $MgCl_2$ 0.01 mol/L; $NaHCO_3$ 0.075 mol/L; KCl 0.0001 mol/L; Na_2SO_4 0.001 mol/L, HEPES 0.01 mol/L, pH 7.5.
   (b) Solution #2—solution#1 diluted ten times with 0.01 mol/L HEPES, pH 7.5.
   (c) The plain tap water acquired in the sufficient amount on the first experimental day.

2. During the first measurement session, the potentiometric data from water samples contaminated with toxic and nontoxic bacterial strains were acquired and employed for construction of a partial least squares-discriminant analysis (PLS-DA) model [15] for classification between two types of samples. This model was tested using leave-one-out validation [16] and has shown 100% specificity and selectivity in toxicity determination; that is, there were no misclassified samples.

3. The toxicity of the samples measured on other experimental sessions was predicted using this PLS-DA model. It was found, however, that the number of misclassified samples in these predictions was significant. The main reason for this discrepancy is the drift of the sensor readings over time. Thus, the numerical values of sensor readings obtained in new samples are not "suitable" for PLS-DA model based on the initial sensor readings. As an example of sensor drift, Fig. 5 shows the differences in sensor readings in standard solution #2 obtained in the first and the second experimental sessions. The particular readings of sensor s_j in standard solutions are shown. In order to correct these drifts the procedure of sensor response standardization described above was employed.

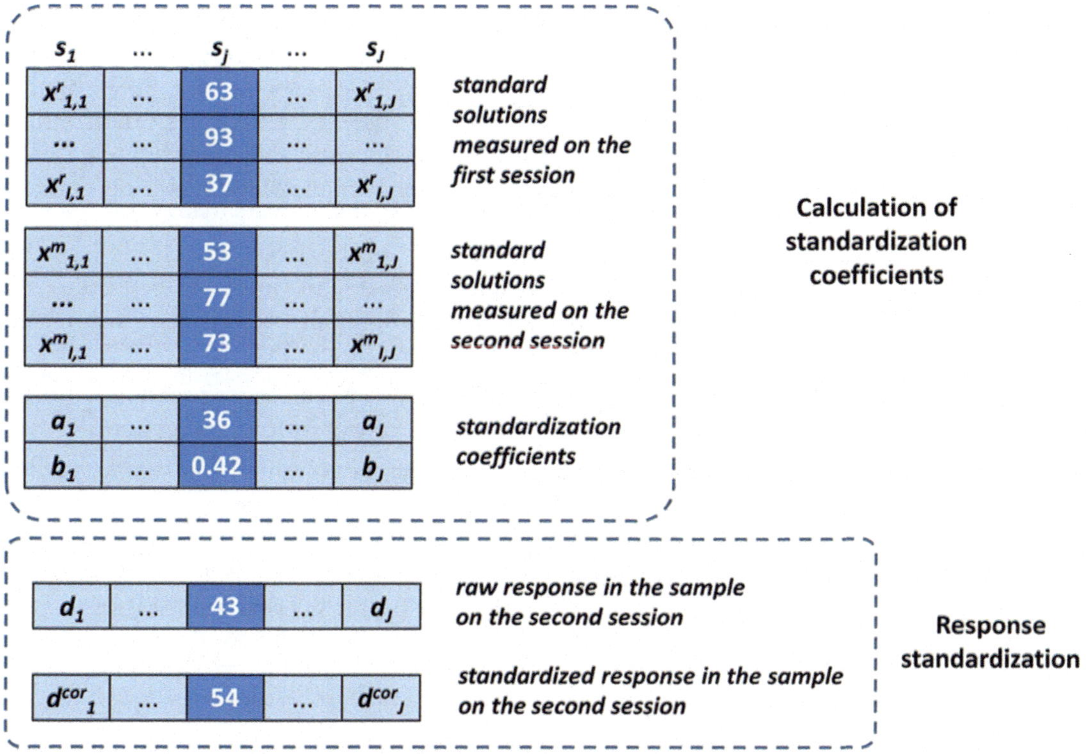

Fig. 5 Standardization of potentiometric sensor array

4. The first experimental session was employed as the "reference day." Following **steps 4–6** from Subheading 3.2, the responses of all sensors in all measurement sessions were standardized. First, standardization coefficients $\boldsymbol{a}$ and $\boldsymbol{b}$ were calculated (particular values are shown for the sensor s_j) using least squares procedure according to the equation shown in Fig. 4:

$$\begin{matrix} \boldsymbol{a}_j \\ \boldsymbol{b}_j \end{matrix} = \left(\boldsymbol{F}^T \boldsymbol{F}\right)^{-1} \boldsymbol{F}^T \boldsymbol{X}_j^r \tag{4}$$

where

$$\boldsymbol{F} = \begin{matrix} 1 & 53 \\ 1 & 77 \\ 1 & 73 \end{matrix}$$

$$\boldsymbol{X}_j^r = \begin{matrix} 63 \\ 93 \\ 37 \end{matrix}$$

5. Second, these coefficients were used to correct the sensor reading in the second measurement session (particular values are shown in Fig. 5):

$$\boldsymbol{d}_j^{cor} = 36 + 0.42 * 43 \tag{5}$$

6. The same procedure must be repeated for all the sensors in all measuring sessions. It is obvious that standardization coefficients are different for different sensors and for different sessions. Implementation of the described standardization procedure affords the use of the multisensor system data with the same mathematical model over an extended period of time.

3.4 Drift Correction Notes

1. It is important to take the appropriate precautions (storage conditions and handling protocol) in order to keep the compositions of all these solutions unaltered through the whole duration of the experiment.
2. The recommended difference between sensor readings should not be smaller than 10 mV (for potentiometric systems).
3. It is recommended to take the measurements in standard solutions on each measurement day (the day when samples of interest are being measured). Take appropriate precautions to ensure the long-time stability of the chemical composition of all standard solutions. Depending on the composition of the solutions these conditions may vary and may include storage at low temperatures, storage in the darkness, storage in inert gas atmosphere, etc.
4. Generally, as the reference day, one can employ the data taken in standard solutions on any day of the measurements. Then all

the rest of the data can be adjusted to this particular day using corresponding correction coefficients.

5. The term "day" is quite arbitrary here, and the measurements can be standardized toward any reasonable stage of the experiment. Depending on the drift intensity the measurements can be standardized between different periods within 1 day, 1 week, 1 month, etc.
6. The increase in the number of standardization solutions will lead to the higher precision in standardization coefficient calculation and to the more precise data correction. However, this increase will lead to the growth of workload and time required for correction; thus a reasonable compromise must be chosen. The minimal recommended number of different standardization solutions is three.
7. Calculation of correction coefficients is based on ordinary least squares; thus any mathematical software capable of performing this procedure can be employed for implementation of the described drift correction protocol.
8. Since the correction coefficients are calculated using linear equation they can be a poor estimate of real drift if the latter is nonlinear in nature. In such cases, more sophisticated procedures may be required in order to compensate for the changes in sensor readings in standard solutions.
9. Obviously, the described procedure is only effective for the properly functioning sensors. If the reason for the observed sensor drift was associated with the sensor breakdown, then no mathematical correction can compensate for that.
10. Periodic measurements in standard solutions are highly recommended for all long-term ET studies. These data will afford the tracking of the changes in sensor performance, allowing one to take appropriate measures to correct possible sensor drifts and to exchange malfunctioning sensors.

Acknowledgments

This work was partially financially supported by Government of Russian Federation, Grant 08-08.

References

1. Ciosek P, Wróblewski W (2007) Sensor arrays for liquid sensing—electronic tongue systems. Analyst 132(10):963–978. https://doi.org/10.1039/b705107g
2. Bratov A, Abramova N, Ipatov A (2010) Recent trends in potentiometric sensor arrays-a review. Anal Chim Acta 678(2):149–159. https://doi.org/10.1016/j.aca.2010.08.035
3. Brereton R, Jansen J, Lopes J, Marini F, Pomerantsev A, Rodionova O, Roger J, Walczak B, Tauler R (2017) Chemometrics in analytical chemistry-part I: history, experimental design and data analysis tools. Anal Bioanal

Chem 409(25):5891–5899. https://doi.org/10.1007/s00216-017-0517-1

4. Analytical Methods Committee (1994) Analyst 119:2363–2366. https://doi.org/10.1039/AN9941902363
5. https://www.fossanalytics.com/en/products/milkoscan-ft1, last accessed on 30 June 2018
6. Oliveri P, Casolino MC, Forina M (2010) Chemometric brains for artificial tongues. Adv Food Nutr Res 61(C):57–117. https://doi.org/10.1016/B978-0-12-374468-5.00002-7
7. Bouveresse E, Massart D (1996) Standardisation of near-infrared spectrometric instruments: a review. Vib Spectrosc 11:3–15. https://doi.org/10.1016/0924-2031(95)00055-0
8. Wang Y, Veltkamp D, Kowalski B (1991) Multivariate instrument standardization. Anal Chem 63:2750–2756. https://doi.org/10.1021/ac00023a016
9. Brown SD (2009) Transfer of multivariate calibration models. In: Brown SD, Tauler R, Walczak B (eds) Comprehensive chemometrics, vol 3. Elsevier B.V., Amsterdam, chapter 3.08, 2009
10. Khaydukova M, Medina-Plaza C, Rodriguez-Mendez M, Panchuk V, Kirsanov D, Legin A (2017) Multivariate calibration transfer between two different types of multisensor systems. Sensors Actuators B 246:994–1000. https://doi.org/10.1016/j.snb.2017.02.099
11. Daszykowski M, Walczak B, Massart D (2002) Representative subset selection. Anal Chim Acta 468:91–103
12. Kennard R, Stone L (1969) Computer aided design of experiments. Technometrics 11:137–148
13. Bro R, Smilde A (2014) Principal component analysis. Anal Methods 6:2812–2831. https://doi.org/10.1039/C3AY41907J
14. Panchuk V, Lvova L, Kirsanov D, Gonçalves CG, Di Natale C, Paolesse R, Legin A (2016) Extending electronic tongue calibration lifetime through mathematical drift correction: case study of microcystin toxicity analysis in waters. Sensors Actuators B 237:962–968
15. Ballabio D, Consonni V (2013) Classification tools in chemistry. Part 1: Linear models. PLS-DA. Anal Methods 5(16):3790–3798
16. Burman P (1989) A comparative study of ordinary cross-validation, v-fold cross-validation and the repeated learning-testing methods. Biometrika 76:503–514

Chapter 16

Computational Modeling for Biomimetic Sensors

Icell M. Sharafeldin, Jessica E. Fitzgerald, Hicham Fenniri, and Nageh K. Allam

Abstract

Computational modeling has become an important tool for scientists to both predict the properties of materials and systems and to gain a better understanding of the underlying mechanisms. This chapter is a brief yet holistic introduction to computational modeling, focusing on density functional theoretical (DFT) methods. The different types of computational modeling methods, including molecular mechanics, semi-empirical, and ab initio methods, as well as the different software available for computational calculations are discussed. A step-by-step guide is presented using Gaussian16 software to introduce the basics of computational modeling based on our work with biomimetic polymer beads. However, the guide presented here is not limited to this particular system; it can be applied to any computational modeling case. The computational modeling methods for the building of the structures are described, and the calculation parameters, such as basis sets and exchange-correlation functionals, are explained. The output data and results are presented and discussed. Two simulation features were the focus of this work: (1) the simulation of the Raman spectra and (2) the different solvation environments. While some researchers in the field believe that computational simulation should be performed before the lab experiments, in fact they should be done simultaneously. This is so that the output of the experimental data can be used as the input of the computational parameters, as a form of semiempirical modeling, in order to achieve more accurate results for predicting the behavior of future experiments and understanding the atomic forces and mechanisms.

Key words Computational modeling, Density functional theory, Barcoded polymer, Multivariate data analysis

1 Introduction

Electronic nose (e-nose) and electronic tongue (e-tongue) circuits analyze the chemical constituents of a gas or liquid, respectively, using arrays of active sensing materials. The efficient active matter necessitates rapid responsivity of the targeted analyte or set of analytes upon exposure, with high sensitivity and selectivity. Scientists carry out extensive laboratory experimentation to design a plethora of active materials for sensing devices for the different types of target analytes. It has recently become widespread practice to use computational modeling to aid in the design of these active

Jessica E. Fitzgerald and Hicham Fenniri (eds.), *Biomimetic Sensing: Methods and Protocols*, Methods in Molecular Biology, vol. 2027, https://doi.org/10.1007/978-1-4939-9616-2_16,

materials for e-nose and e-tongue sensors. Computational modeling of different materials can be used to predict the response of the materials to the targeted gases or liquids and gain an in-depth understanding of the sensing materials' properties. These data can be used to select the most promising sensing materials for the targeted analyte(s) prior to their synthesis and testing in a laboratory environment.

Molecular mechanics, semiempirical, and ab initio computational modeling methods are used to define a molecular system and predict its optimum geometry, stability, and electronic states among other properties [1]. The first, molecular mechanics, is an empirical classical mechanics-based method [2], where a molecule is described as a collection of particles connected by a molecular force field composed of harmonic or elastic forces. These methods are relatively inexpensive and fast, and provide a high degree of accuracy for large molecular systems consisting of thousands to millions of atoms [3]. The latter two, semiempirical and ab initio methods, on the other hand, are based on quantum mechanical methods and are restricted to smaller systems, which provide a higher accuracy than molecular mechanical methods. Semiempirical methods differ from ab initio methods due to their utilization of empirical data, to allow for the inclusion of electron correlation effects [4]. Ab initio methods, however, consider the electron correlation effects computationally, resulting in the highest computational costs among the three types of computational methodologies [5]. Ab initio calculations rely solely on quantum mechanics and attempt to solve the Schrödinger equation using basic physical constants such as speed of light and masses of atomic particles [6].

Density functional theory (DFT) methods, used herein, are considered ab initio methods, offering high accuracy in determining the system energies, and are not limited to element type, class, or chemical environment [7]. The quantum mechanical wave function contains all the information about a given system: its structural composition, lattice parameters, magnetic and electronic properties, as well as its defects [8]. The Schrödinger equation can be solved exactly for a simple 2-D square potential or a hydrogen atom to derive the wave function and determine the allowed energy states of the system. However, for many-body systems, scientists and mathematicians have yet to find an exact solution to the Schrödinger equation. DFT allows for the approximation of the solution to the Schrödinger equation for these many-body systems with high accuracy. DFT is derived from the Hohenberg-Kohn set of approximations; the first approximation states that the sum of the kinetic and electron-electron repulsion energies of an N-electron system is a universal functional of its density, while the second approximation assumes that if one knew the universal functional of the density of electrons, then one can solve for the energy by utilizing the variational principle [9]. To this end, many software programs have been developed to employ the DFT method to solve the Schrödinger equation and define the properties of different molecules and structures. Each software package has its own set of library pseudopotentials, focus materials, type of functionals, and other parameters. Some examples will be introduced here, such as CASTEP, VASP, and GAUSSIAN; however, a more complete list of software is available online [10].

CASTEP was used previously in our work to investigate the adsorption mechanisms of CO on several copper facets [11] and the charge transfer between NO_2 and metal-doped carbon nanotubes [12] as it uses a plane wave basis set to calculate the electronic properties of crystalline solids, surfaces, molecules, liquids, and amorphous materials [13]. CASTEP permits geometry optimization and finite temperature molecular dynamics with implicit symmetry, periodicity in structure, and geometry constraints. VASP calculates quantum mechanical molecular dynamics (MD) using either Vanderbilt pseudopotentials or projector-augmented wave method, and a plane wave basis set using the projector-augmented wave method to describe the electron-core interaction [14]. The code also allows for the use of post-DFT corrections such as hybrid functional mixing DFT and Hartree-Fock exchange, many-body perturbation theory (the GW method), and dynamic electronic correlations within the random-phase approximation that offers higher accuracy [14]. GAUSSIAN16, the focus software of this chapter, was developed by John Pople [15] to speed up and improve the performance of the Hartree-Fock calculations by using Gaussian orbitals instead of the initial Slater-type orbitals. GAUSSIAN16 models can be applied to both known and stable species as well as compounds that are difficult to observe experimentally, whether due to their toxic, radioactive, or combustible nature or transition structures in chemical reactions [15].

Selecting the appropriate software to use can be challenging and depends on several factors, including its availability and the computational resources at hand. To select a software, one must first define the system or structure under investigation and then conduct a thorough literature review of software used to calculate its properties or to investigate similar structures. The accuracy of the results must also be taken into consideration, as well as the time allotted for the investigation and the computational cost, as higher accuracy comes with a higher computational cost.

Crescenzi et al. used computational methods to predict the technological potential of 5,6-dihydroxyindole (DHI)-based porphyrin-type tetramer for eumelanin biopolymers. Their computations were performed either in vacuo or by adoption of a polarizable continuum model (PCM) to account for the influence of the solution environment [16]. The Gaussian package was used to determine the electronic properties of the tetramer to investigate how oxidation affected the stability of the structure as well as the

HOMO-LUMO energy levels. DFT simulations may be used as a baseline for the design of innovative bioinspired optoelectronic materials. Comba et al. [17] investigated the three pathways for the reaction of bispyridine-iron(II) complexes with H_2O_2. In this case, DFT calculations revealed a direct conversion of the bispyridine-iron(II)-hydrogen peroxide complex to the corresponding iron(IV)-dihydroxo compound. This compound is a possible oxidant for olefin dihydroxylation. The calculations were also able to predict the frequency spectra of the olefins to verify that they are in fact the minima on the potential energy scans and determine the zero-point energy corrections.

As can be seen from the examples above, there are many properties that can be defined using computational modeling, and these are by no means exhaustive. Many other features can also be analyzed, such as optimum geometrical structures, Raman and IR vibrational energy levels, NMR spectra, UV and visible electronic energy levels, thermochemical properties, potential energy surfaces showing barrier heights and possible transition states, electron affinities, dipole moments, polarizabilities, electron density maps and population analyses, and magnetic shielding tensors. The aim of this chapter is to present a practical example of computational modeling for biomimetic sensors. Specifically, a practical example is presented based on our work [18] on a spectroscopic chemical sensor of an array of nine self-encoded polymer beads. Experimental methods of vapor and liquid sensing of these beads are outlined in Chapter 1. The Raman spectra before and after exposure to organic volatiles create a unique vibrational fingerprint for each of the polymer beads. We will show how GAUSSIAN16 can be used to investigate the structures as well as to obtain the Raman spectra.

2 Polymer-Based Biomimetic Sensors

The first part of this section focuses on building molecular structures (Subheading 2.1), followed by calculating the optimized properties at the ground state (Subheading 2.2). Third, we then show how to perform Raman calculations and compare them to the available spectra in the literature (Subheading 2.3). The fourth section (Subheading 2.4) explores testing in different solvation environments, and the final section (Subheading 2.5) illustrates other important properties that DFT and computational modeling can provide.

2.1 Building the Structure

The first step to any simulation is building the molecule or structure under investigation. GAUSSIAN16, as well as other software packages such as CHEMDRAW, allows the users to build any structure. The structure should be drawn as close to experimental angles and bond lengths as possible. There are also ready-made files

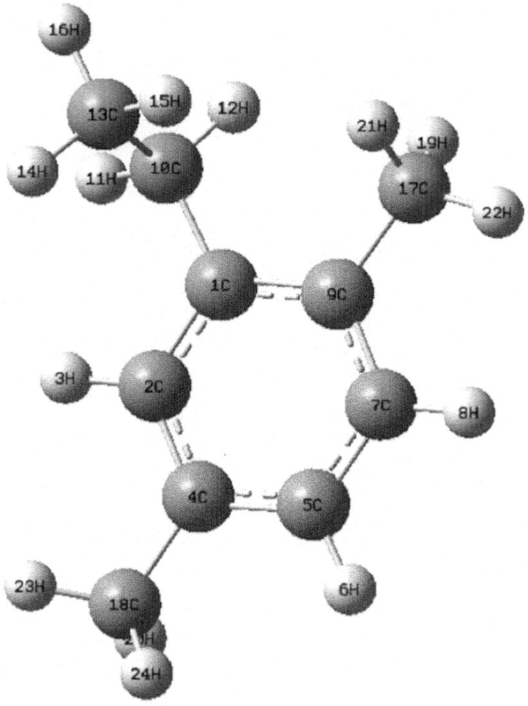

Fig. 1 Structure of 2,5-dimethyl styrene monomer

available to download, but these files may not always describe the exact structure under investigation. In this chapter, the 2,5-dimethyl styrene structure (Fig. 1) was drawn and then ***geometry optimization*** calculations were run to generate a structure with optimal bond lengths and angles. The 2,5-dimethyl styrene structure is modeled without the 10C–13C double bond to simulate its condition under the case of polymerization.

2.2 Geometry Optimization and Selecting Parameters

The optimal configuration of any molecular structure is the lowest energy-equilibrium (ground-state) structure. Any other state will produce different surface and binding energies as well as structural and electrical properties. DFT runs geometry optimization calculations and finds the optimal structure by locating a stationary point on its potential energy surface of the structure using the first derivative of the energy with respect to the nuclear position.

1. On ***GaussView***, the graphic user interface for GAUSSIAN16, select "*Calculate*" from the main menu and select the "*Gaussian Calculation Setup*" option. The pop-up window provides a set of parameters for the calculation configuration.
2. In the "*Job Type*" tab, select the "*Optimization*" option and move on to the "*Method*" tab where the main decisions must be made concerning the calculation methodology and accuracy.

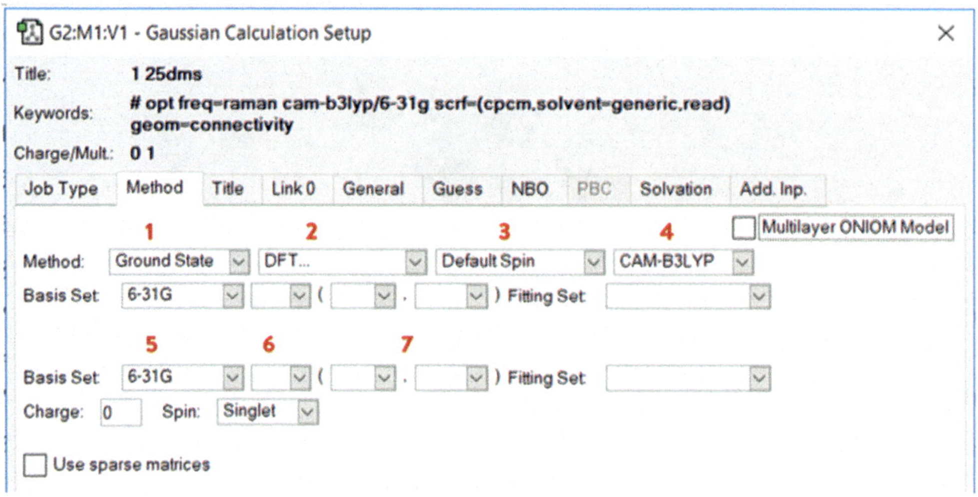

Fig. 2 Selecting the computational criteria from GaussView "Method" tab

The main options are the *Method*, and *Basis Set* as shown in Fig. 2.

3. For this work, we selected "*Ground State*" for option 1 to find the structure at the lowest energy equilibrium and "*DFT*" as the preferred method in option 2. The rest of the options are parameters that are varied to find the optimum combination resulting in the highest correspondence with experimental findings (*see* **Note 1**).
4. **Option 3—Spin:** There are four options to the spin settings in GAUSSIAN16: *Default Spin*, *Restricted*, *Unrestricted*, and *Restricted-Open*.
   (a) "*Default*" sets calculations to unrestricted only if the spin multiplicity of the structure is higher than 1.
   (b) "*Restricted*" calculations are for systems of paired electrons, i.e., multiplicity is 1, and each orbital is fully occupied with two electrons (closed shell).
   (c) "*Unrestricted*" calculations are for systems of multiplicity higher than 1; that is, the system has unpaired electrons.
   (d) "*Restricted open-shell*" calculations force the inner orbitals that are fully occupied with two electrons to be calculated as restricted and the orbitals containing one electron as unrestricted, which gives higher accuracy at higher computational cost [19].
5. **Option 4—Exchange-Correlation Functionals:** Approximate solution functionals based on the electron density that describe the kinetic and electron-electron interaction energies. There are two main classes of approximations, the local density

approximations (LDA) [20] and the generalized gradient approximations (GGA) [21].

(a) The LDA approximations depend on the electron density at any point in space and adopt the homogenous electron gas model where this density value is assumed to be uniform. LDA approximations can describe the atomic structure and elastic and vibrational properties accurately, yet cannot adequately describe the energetics of chemical reactions due to an overestimation of the binding energies [22].

(b) The GGA approximations incorporate the effects of electron density inhomogeneities by including the gradient of the electron density, thereby improving the description of the energy barriers and binding energies of systems [23].

6. **Option 5—Basis Sets:** A collection of functions that define the volume in which the equation is solved. A basis set is a set of functions, either atomic orbitals centered on atoms or plane waves that can be combined linearly to create molecular orbitals [24].
7. **Option 6—Diffuse Functions:** An extension of the Gaussian basis functions that can be added to allow the "*tail*" portion of the atomic orbitals to be more flexible at greater distances away from **the** nucleus. Diffuse basis functions are used to describe anions and dipole moments as well as intra- and intermolecular bonding. One plus sign indicates that the diffuse functions are on the heavy atoms and two plus signs indicate that diffuse functions are also added to light atoms such as hydrogen and helium [25].
8. **Option 7—Polarization Functions:** This can be added to the minimal basis set to obtain a better description of the polarization of the atomic electron density in molecule. This allows the molecular orbitals to become asymmetrical, if needed. This is important for chemical bonding, where the adsorption causes the molecular orbitals to become polarized [25]. One asterisk means that polarization functions are added to heavy atoms, and two asterisks indicate that polarization functions are also added to light atoms, hydrogen and helium. The 3-21G** is synonymous to 3-21G(d,p).
9. To start off the **calculation**, select any variable to test for which the run is executed without any complications, preferably "LSDA" and "STO-3G" as this selection results in a relatively short calculation time.
10. After selecting the starting parameters, click "*Submit*" on the "*Gaussian* **Calculation** *Setup*" menu to start the calculation.

2.3 Interpreting Optimization Results

A successful run will produce a *.log* file.

1. Open the *.log* file using notepad to ensure that the "*normal termination*" text is present and that the run did not result in erroneous output.
2. To view a summary of some of the important results, open the *.log* file in ***GaussView***, right click on the window, and then select "*Results*," and then "*Summary*." An example of the "*Summary*" window is presented in Fig. 3. The window provides a summary of the calculation parameters selected such as type, method, basis set, charge, and spin.
3. The total energy of the molecular system is presented in atomic units (a.u.). The atomic units are in Hartrees, 1 Hartree ~2625.5 kJ/mol ~27.21 eV [6]. The energy term here represents the energy due to the electronic motion and the nuclear-nuclear repulsion of the atoms in the structure. This value, however, does not include the thermal energy component. This thermal correction is found in the *.log* output file under the "*Thermochemistry*" section titled "*Thermal Correction to Energy*" (value highlighted in Fig. 4). This value is then added to the total energy to find the real total energy of the system.
4. The imaginary frequency found is zero, which signifies that the true minimum of the potential energy surface was found,

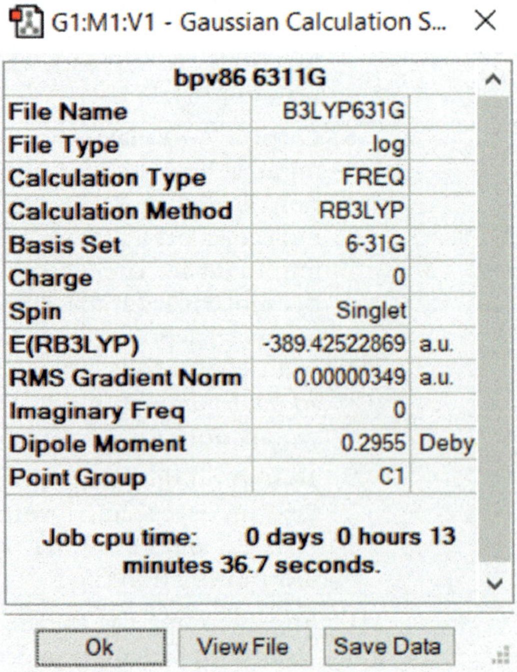

Fig. 3 "Summary" window

```
-------------------
- Thermochemistry -
-------------------

Zero-point correction=                          0.214403 (Hartree/Particle)
Thermal correction to Energy=                   0.225039
Thermal correction to Enthalpy=                 0.225983
Thermal correction to Gibbs Free Energy=        0.177592
```

Fig. 4 Thermal correction value in **.log** file

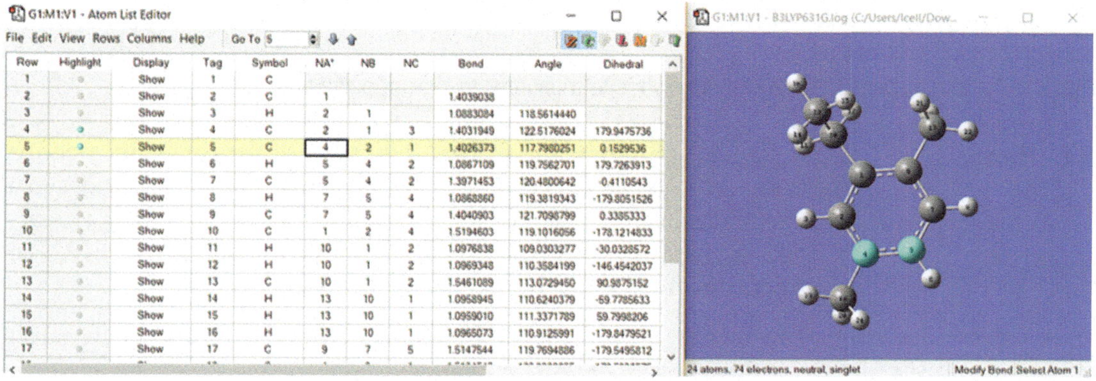

Row	Highlight	Display	Tag	Symbol	NA*	NB	NC	Bond	Angle	Dihedral
1		Show	1	C						
2		Show	2	C	1			1.4039038		
3		Show	3	H	2	1		1.0883084	118.5614440	
4		Show	4	C	2	1	3	1.4031949	122.5176024	179.9475736
5		Show	5	C	4	2	1	1.4026373	117.7980251	0.1529536
6		Show	6	H	5	4	2	1.0867109	119.7562701	179.7263913
7		Show	7	C	5	4	2	1.3971453	120.4800642	-0.4110543
8		Show	8	H	7	5	4	1.0868860	119.3819343	-179.8051526
9		Show	9	C	7	5	4	1.4040903	121.7098799	0.3385333
10		Show	10	C	1	2	4	1.5194603	119.1016056	-178.1214833
11		Show	11	H	10	1	2	1.0976838	109.0303277	-30.0328572
12		Show	12	H	10	1	2	1.0969348	110.3584199	-146.4542037
13		Show	13	C	10	1	2	1.5461089	113.0729450	90.9875152
14		Show	14	H	13	10	1	1.0958945	110.6240379	-59.7785633
15		Show	15	H	13	10	1	1.0959010	111.3371789	59.7998206
16		Show	16	H	13	10	1	1.0965073	110.9125991	-179.8479521
17		Show	17	C	9	7	5	1.5147544	119.7694886	-179.5495812

Fig. 5 Results of the obtained bond length and bond angle for the optimized structure

ensuring that the resulting structure is, in fact, the optimal configuration for the atoms.

5. The magnitude of the dipole moment in Debye, a measure of the system's asymmetry, is calculated and the dipole vector can be presented on the structure by right clicking and then selecting "*Results*" and "*Charge distribution*" and then from the "*Dipole moment*" section check the "*Show Vector*" box.

6. To view the bond angles and lengths for the structure, right click and select "*Edit*," and then "*Atom list*." An example of the structural geometry values is presented in Fig. 5 along with the corresponding molecule structure. Each atom is given a "Tag" identification number in the structure. The "Bond" is the length between the atom selected and the atom in the "NA" column. The "Angle" is the angle determined by the three atoms of "NA," "NB," and the selected atom. The "dihedral" is the dihedral angle formed by the intersecting planes defined by the four atoms under "NA," "NB," "NC," and the selected atom. After successful optimization of the structure, single-point energy calculations can be run to find more properties such as Raman, IR, and charge distribution.

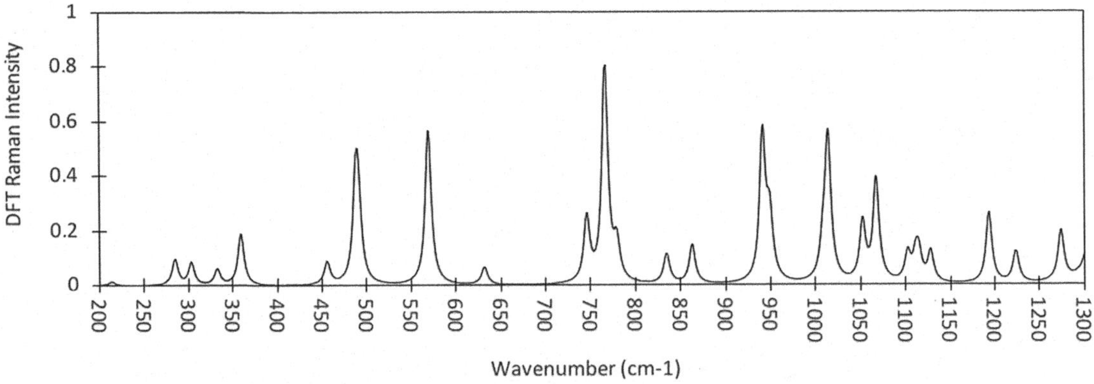

Fig. 6 DFT Raman spectra of 2,5-dimethyl styrene monomer as part of polymer

2.4 Raman

A Raman spectrum, the second derivative of the electronic energy, is a spectroscopic property used to study the vibrational modes in a system and can be used to identify unique substances from characteristic pattern, also known as "fingerprinting." GAUSSIAN16 calculates the IR and Raman frequencies in two ways:

1. In the "Job Type" tab, select "Opt+Freq," and then select "yes" to "*Compute Raman.*"
2. After first finishing the optimization run, start another run, but in this case select "*Frequency*" in the "*Job Type*" tab and then select "*yes*" to the "*Compute Raman.*"
3. After submitting and completing the run successfully, right click on the output file, select "*Result*" and then "Vibrations," and then click "*Spectrum*" on the "*Vibrational Spectra*" window. This window will display the IR, Raman, P-depolarization, and U-depolarization spectra (*see* **Note 2**). Figure 6 shows the DFT-derived Raman in vacuo for the 2,5-dimethyl styrene described above. The agreement between DFT and experimental Raman spectra is good in terms of peak positions and pattern, though it is less accurate in intensity. However, a more quantitative means of comparing the spectra was used to compare the DFT and experimental Raman spectra (*see* **Note 3**).

2.5 Modeling Solvation Environments

There are several ways of modeling the solute in solution. GAUSSIAN16 employs the self-consistent field reaction (SCRF) solvation method using the conductor-like polarizable continuum model (CPCM). The CPCM theoretically places the solute in a cavity within a continuous dielectric medium, which simulates the solvent properties [26] (*see* **Note 4**).

1. To simulate solvation effects in GAUSSIAN16, select the "*Solvation*" tab from the "*Gaussian Calculation Setup*" pop-up window.

Table 1
Summary of solvent dielectric constants used and resulting changes in average parameters

Solvent	Dielectric constant	Δ Bond length (Å)	Δ Bond angle (°)	Δ Dihedral angle (°)
Acetone	21.83	0.0004781	-0.0260	-0.3109
Carbon disulfide	2.60	0.0003812	5.0393	-0.2418
Cumene	2.37	0.0003721	5.0395	-0.2382
Propyl benzene	2.36	0.0003721	5.0395	-0.2382
Pentane	1.80	0.0003372	5.0404	-0.2249

2. Then use the default "*Model*" type unless otherwise specified or necessary.
3. Select the solvent from the drop-down menu (*see* **Note 5** if the solvent is not on the list).
4. Submit the job for running.

   After the completion of a successful run, one can view the summary to find the structural changes of the molecule. Table 1 presents the solvents and their corresponding dielectrics used in our example, and the resulting energy difference and average difference in bond length, bond angle, and dihedral angle of the 2,5-dimethylstyrene monomer upon exposure to the different analytes. The overall effect of analyte vapor exposure on structural geometry is minimal, and the calculated geometrical parameters represent a good approximation and were the bases for calculating the vibrational frequencies (*see* **Note 6**).

   In our example, adding solvents to the polymer beads induced a change in the properties, thereby changing the Raman spectrum of each polymer bead upon interaction. The solvents used for this work were acetone, propyl benzene, pentane, carbon disulfide, and pentane. The angles (θ), which determine the degree of change of the Raman spectra of the polymer beads upon solvent exposure were calculated and compared with the experimental counterparts. The results are presented in Fig. 7 and are in good agreement with the experimental values (*see* **Note 7**).

2.6 Molecular Orbitals, Energy Levels, HOMO-LUMO, and Band Gaps

Other important features of DFT modeling that aid in understanding the molecular composition and adsorption properties are the molecular orbital energy states, HOMO-LUMO levels, and band gap. A molecular orbital (MO) is a mathematical function describing the wavelike behavior of an electron in a molecule. This function can be used to calculate chemical and physical properties such as the probability of finding an electron in any specific region.

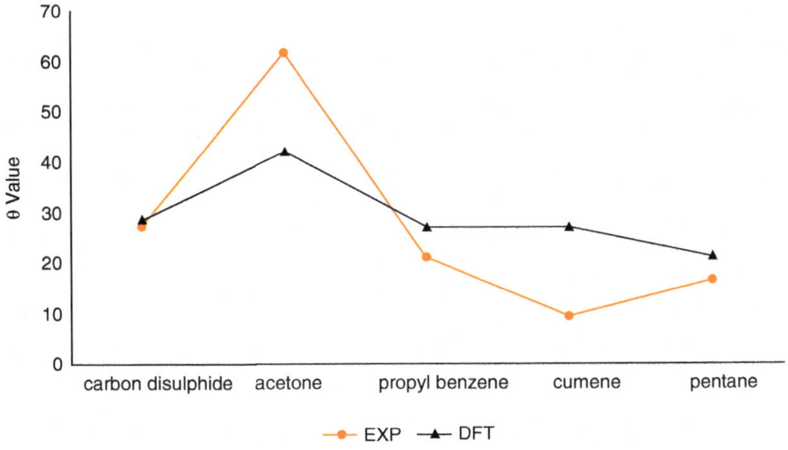

Fig. 7 Correlation of θ between experimental and DFT-derived Raman spectra

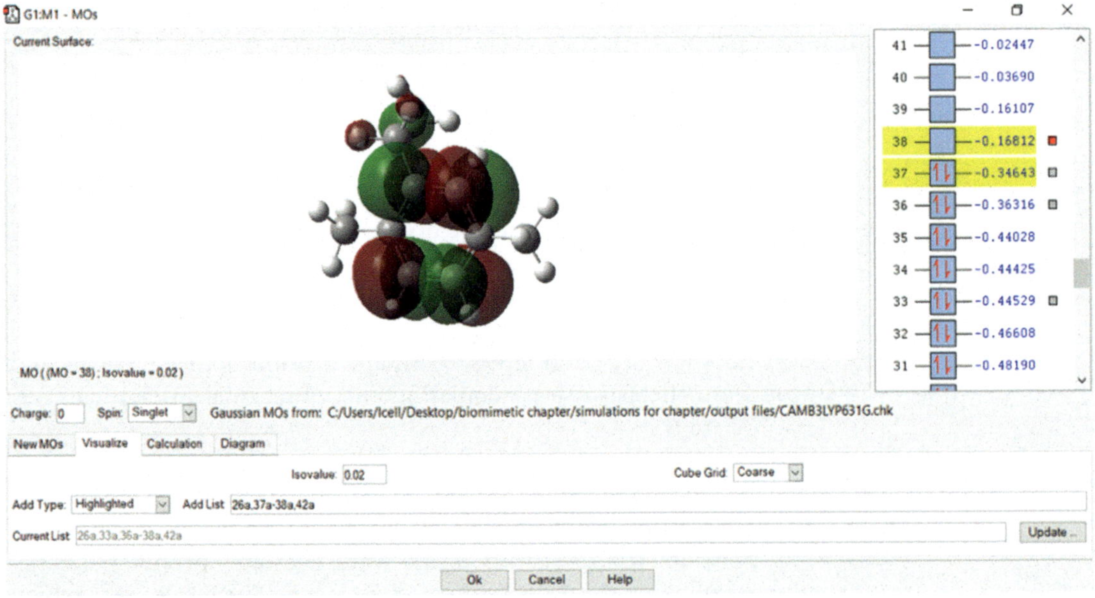

Fig. 8 Molecular orbital window

1. To generate molecular orbitals in GAUSSIAN16, right click on the output file opened in ***GaussView***, and then select "*Edit*" and then "*MOs*."
2. The MO window will open as shown in Fig. 8. The right panel is a list of all the orbitals in the structures and their corresponding energy values as well as their occupancy. To view the molecular orbital surface, one must first generate the file.
3. Select the "*New MOs*" tab, and if there is a check point file generated, click "*Load*." If the check point file does not exist,

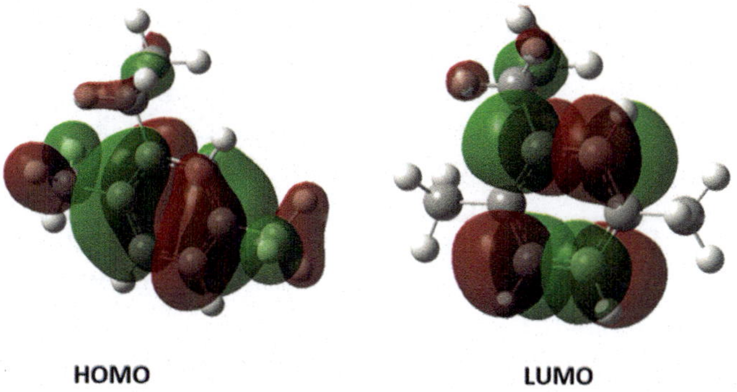

Fig. 9 HOMO and LUMO molecular orbital surfaces using GAUSSIAN16

select "*Generate*" to perform a calculation where the molecular orbitals are computed.

4. One can then highlight any orbital and select the "*Visualize*" tab and select "*Update*" to view the surface.

Presented in Fig. 9 are the HOMO and LUMO levels for 2,5-dimethylstyrene, which are highlighted in yellow in Fig. 8. The HOMO and LUMO are termed the frontier orbitals, and the energy gap between them is used to predict the strength and stability of transition metal complexes as well as their optical properties.

3 Conclusions

In this chapter, we give a brief description of computational modeling and how it can be used to model a variety of systems with a focus on polymer-based structures for biomimetic sensors. As seen, the advantages of using computational modeling for this purpose include (a) faster screening of potential materials for initial synthesis and experimentation; (b) lower consumption of laboratory materials and resources; (c) analysis at a relatively low cost compared to that of laboratory instrumentation and experimental supplies; (d) exploration of new parameters of the material such as optical properties; (e) a fundamental understanding of microscopic chemical reactions and bonding mechanisms, which can then be compared with experimental results to prove effectiveness; and (f) ability to simulate durability in extreme conditions.

We focused on using the GAUSSIAN16 software because of its availability, accuracy, and flexibility; however, computational modeling can be conducted using a wide range of software, each offering its own advantages and disadvantages. Laboratory experimentation is a vital part of any research to verify theoretical understanding; however, one of the most valuable advantages of using computational modeling is that it both saves laboratory materials and resources and helps researchers gain a better understanding of theoretical mechanisms. Biomimetic sensing studies can especially benefit from a deeper understanding of sensors' adsorption mechanisms, as they make use of a wide range of materials and require high screening of possible materials. This would help improve the speed of development of highly selective and sensitive sensor arrays for several vital applications such as biomedical research, environmental monitoring, and health and safety.

4 Notes

1. The goal is to find the best combination of these variables to produce results as close to the experimental values as possible. One way to determine the combination is to go through a literature review on DFT work of similar structures to find a combination to serve as a starting point.
2. The spectra can be copied and pasted in a separate Excel file for your own plotting convenience.
3. The correlation can be quantified in several ways; in our work we used principal component analysis (PCA) to efficiently capture the variances. The Raman spectra are represented as a vector in n-dimensional space where each dimension represents a relative intensity for any given wavenumber. To compare the experimental to the DFT Raman spectra, denote two vectors, y_i and y_i', to represent each Raman spectrum, respectively. We calculated the θ value, which is defined as the angle between the experimental (y_i) and DFT (y_i') vectors for the sensing material of interest. The angle is determined by the inner product of the two vectors:

$$\cos\,\theta = \frac{|y_i * y_i'|}{|y_i||y_i'|} \tag{1}$$

   From Eq. 1, if $y_i = y_i'$, then $\cos\,\theta = 1$, and thus $\theta = 0$. The angle reflects the degree of difference between the experimental and DFT Raman spectra. Therefore, the smaller the angle between y_i and y_i', the closer the calculated Raman is to the experimental spectra.
4. Hence, the dielectric constant used for the solvents has a significant effect on the results of the simulation. The charge distribution of the solute polarizes the solvent, which in turn polarizes the solute until an equilibrium is reached. This mutual polarization derives the polarizable continuum model (PCM). The role of the cavity is to isolate the solute from the solution, thereby simplifying the calculations by only considering the apparent charges on the cavity surface from both ends.

There are different cavity shapes that can be used to isolate the solute. The simplest is a spherical cavity surrounding the solute; one can also model a surface of constant electron density and a solvent-accessible surface. The GAUSSIAN16 PCM model uses the superposition of interlocked spheres centered on the atoms defined by the van der Waals values; it is a good trade-off between cost and accuracy [6]. The default GAUSSIAN16 solvation model is IEFPCM, which uses a relatively accurate model of the charge decay of the molecules experienced beyond the cavity that were disregarded in former versions [6].

5. If the solvent is not on the list, select "*Other*" and then type "*generic*" in the neighboring space and make sure that the "*Read additional input*" box is checked. Select the "*Add. Inp.*" tab to define the parameters of the solvent to be investigated. Determine the solvents' stoichiometry, solvent name, static (*eps*), and dynamic (*epsinf*) dielectric constant to be added as text in the "*Add. Inp.*" tab. An example of the text to be entered is shown below:

   stoichiometry = C_9H_{12},

   solventname = *cumene*,

   eps = *2.3712*,

   epsinf = *1.758*.

6. In our example, the DFT calculations demonstrate that analyte exposure increases the bond length and dihedral angle in correlation with the dielectric constant of the solvent. The increase is relatively minor for bond length, 10^{-4} Å. Exposure to acetone decreased the bond angle by 0.05°, while all other solvents show an average increase of 5°.

7. The results are in good agreement with the experimental values and the general trend can be observed. Therefore, Raman can be used to identify the target solvent and predict its Raman spectra using computational DFT methods. This can further be improved by repeating the experimental readings, by continuing to explore the different variable combinations, and polymer lengths to improve the correlation.

References

1. Mohamadi F, Richards NGJ, Guida WC et al (1990) Macromodel-an integrated software system for modeling organic and bioorganic molecules using molecular mechanics. J Comput Chem 11:440–467. https://doi.org/10.1002/jcc.540110405
2. Engler EM, Andose JD, Schleyer PVR (1973) Critical evaluation of molecular mechanics. J Am Chem Soc 95:8005–8025. https://doi.org/10.1021/ja00805a012
3. Burkert U, Norman A (1982) Molecular mechanics. Am Chem Soc 177. https://doi.org/10.1002/jcc.540040420
4. Ögretir C, Csizmadia IG (1991) Computational advances in organic chemistry: molecular structure and reactivity. Springer, Netherlands, Dordrecht

5. Szabo A, Ostlund NS (1989) Modern quantum chemistry: introduction to advanced electronic structure theory. McGrw-Hill, New York, NY
6. Foresman JB, Frisch A (2015) Exploring chemistry with electronic structure methods, 3rd edn. Gaussian, Inc., Wallingford, CT
7. Robert P, Yang W (1995) Density-functional theory of atoms and molecules. Oxford University Press, Oxford
8. Segall MD, Lindan PJD, Probert MJ et al (2002) First-principles simulation: ideas, illustrations and the CASTEP code. J Phys Condens Matter 14:2717–2744. https://doi.org/10.1088/0953-8984/14/11/301
9. Yu HS, Li SL, Truhlar DG (2016) Perspective: Kohn-Sham density functional theory descending a staircase. J Chem Phys 145:130901. https://doi.org/10.1063/1.4963168
10. From Wikipedia the Free Encyclopedia (2018) List of quantum chemistry and solid-state physics software. https://en.wikipedia.org/wiki/List_of_quantum_chemistry_and_solid-state_physics_software
11. Gameel KM, Sharafeldin IM, Abourayya AU et al (2018) Unveiling CO adsorption on Cu surfaces: new insights from molecular orbital principles. Phys Chem Chem Phys 20:25892–25900. https://doi.org/10.1039/C8CP04253E
12. Sharafeldin IM, Allam NK (2017) DFT insights into the electronic properties and adsorption of NO_2 on metal-doped carbon nanotubes for gas sensing applications. New J Chem 41:14936–14944. https://doi.org/10.1039/C7NJ03109B
13. Clark SJ, Segall MD, Pickard CJ et al (2005) First principles methods using CASTEP. Zeitschrift für Krist 220:567–570. https://doi.org/10.1524/zkri.220.5.567.65075
14. Hafner J (2008) Ab-initio simulations of materials using VASP: density-functional theory and beyond. J Comput Chem 29:2044–2078. https://doi.org/10.1002/jcc.21057
15. Hehre WJ, Ditchfield R, Pople JA (1972) Self—consistent molecular orbital methods. XII. Further extensions of Gaussian—type basis sets for use in molecular orbital studies of organic molecules. J Chem Phys 56:2257–2261. https://doi.org/10.1063/1.1677527
16. Crescenzi O, D'Ischia M, Napolitano A (2017) Kaxiras's porphyrin: DFT modeling of redox-tuned optical and electronic properties in a theoretically designed catechol-based bioinspired platform. Biomimetics 2:21. https://doi.org/10.3390/biomimetics2040021
17. Comba P, Rajaraman G, Rohwer H (2007) A density functional theory study of the reaction of the biomimetic Iron(II) complex of a tetradentate bispidine ligand with H_2O_2. Inorg Chem 46:3826–3838. https://doi.org/10.1021/ic061129y
18. Fitzgerald J, Sharafeldin I, Bravo-Vasquez JP, et al Microsensor based on cross-reactive, self-encoded polymer beads array
19. Jacob CR, Reiher M (2012) Spin in density-functional theory. Int J Quantum Chem 112:3661–3684. https://doi.org/10.1002/qua.24309
20. Jones RO, Gunnarsson O (1989) The density functional formalism, its applications and prospects. Rev Mod Phys 61:689–746. https://doi.org/10.1103/RevModPhys.61.689
21. Perdew JP, Wang Y (1992) Pair-distribution function and its coupling-constant average for the spin-polarized electron gas. Phys Rev B 46:12947–12954. https://doi.org/10.1103/PhysRevB.46.12947
22. Grossman JC, Mitas L, Raghavachari K (1995) Structure and stability of molecular carbon: importance of electron correlation. Phys Rev Lett 75:3870–3873. https://doi.org/10.1103/PhysRevLett.75.3870
23. Kohn W, Becke AD, Parr RG (1996) Density functional theory of electronic structure. J Phys Chem 100:12974–12980. https://doi.org/10.1021/jp960669l
24. Finocchi F (2011) Density functional theory for beginners-basic principles and practical approaches. Institut des NanoSciences de Paris (INSP) CNRS and University Pierre et Marie Curie, Paris
25. Lewars EG (2016) Computational chemistry-introduction to the theory and applications of molecular and quantum mechanics. Springer, Switzerland
26. Tomasi J, Mennucci B, Cammi R (2005) Quantum mechanical continuum solvation models. Chem Rev 105:2999–3094. https://doi.org/10.1021/cr9904009

INDEX

Jessica E. Fitzgerald and Hicham Fenniri (eds.), *Biomimetic Sensing: Methods and Protocols*, Methods in Molecular Biology, vol. 2027, https://doi.org/10.1007/978-1-4939-9616-2, © Springer Science+Business Media, LLC, part of Springer Nature 2019

Index

Printed by Printforce, the Netherlands